NELSON'S AGRICULTURE SERIES

GENERAL EDITORS:

SIR WILLIAM G. OGG M.A. PH.D. AND G. V. JACKS M.A. B.SC.

SOIL

SOIL

G. V. JACKS
Director, Commonwealth Bureau of Soil Science, Rothamsted

THOMAS NELSON AND SONS LTD
LONDON EDINBURGH PARIS MELBOURNE JOHANNESBURG
TORONTO AND NEW YORK

THOMAS NELSON AND SONS LTD
Parkside Works Edinburgh 9
36 Park Street London W1
117 Latrobe Street Melbourne C1

THOMAS NELSON AND SONS (AFRICA) (Pty) LTD
P.O. Box 9881 Johannesburg

THOMAS NELSON AND SONS (CANADA) LTD
91–93 Wellington Street West   Toronto 1

THOMAS NELSON AND SONS
18 East 41st Street New York 17, N.Y.

SOCIÉTÉ FRANÇAISE D'ÉDITIONS NELSON
97 rue Monge Paris 5

———

First published January 1954
Reprinted 1956, 1958, 1959, 1963

63/   8113

PREFACE

THE purpose of this book is to give the farmer, the student of agriculture and all who realise the importance of the cultivation of the land an insight into the new science of soil management. For, although millions of different soils occur in nature, not one of them will produce the cultivated plants of agriculture without constant care by man. Much of the knowledge of soil fertility has been discovered within the last fifty years, and only by applying this knowledge can the practical man reach the greatest efficiency in a sphere whose immense importance is now recognised as demanding urgent attention.

The subject is an intricate one in which the physics, chemistry and biology of the soil all play their part. My aim has been to place before the reader the latest findings on each aspect of the subject in as simple a manner as possible without diverging too far from the strict principles of scientific orthodoxy.

CONTENTS

vii

CONTENTS

PLATES

SOIL

CHAPTER I

WHAT IS SOIL ?

MANY definitions of soil have been made, some very long, some short. The longer they are the less they define and the more they describe. My definition is very short, very undescriptive and, like most definitions, imperfect. It is— soil is what plants grow in. This includes, and should include, such habitats as the surface layers of a pond in which duck-weed is growing as well as the mud at the bottom of the pond in which the deeper roots of aquatic plants live. It also includes a solution of plant food contained in a bottle or a tank ; indeed, such solutions are among the most perfect of all soils for the production of some economic plants, for example tomatoes, and enormous crops can be obtained under ideal conditions because we can exercise complete control over the composition of the soil. The two parts of any soil that are indispensable to plant growth are the aqueous solution from which plant roots absorb their food, and air to enable plant roots to breathe. Anything which contains air, water and plant food in a state in which plants can get them conforms to our definition of soil.

The soils which cause so much trouble to farmers, gardeners and foresters, however, are much more complex than the systems represented by aerated solutions of plant nutrients. They were created to perform a much wider range of functions than can be performed by water solutions, which usually can grow only one crop and are then exhausted. Most ordinary soils can grow a great variety of plants equally well, and they can go on growing them indefinitely. Indeed, the fertility or plant-producing power of a soil is maintained and often enhanced in nature merely by growing plants. It is the life in soil that gives it its fertility, the characteristic that distinguishes a mineral soil from the mineral particles of which it is mainly composed. The reason why the fertility of a soil consisting of a solution of plant food is quite evanescent is that there is no mechanism for replacing the plant food absorbed by growing plants. But if the plant material is allowed to fall back into the water when it dies, and is then suitably decomposed by living organisms, thus releasing the absorbed plant nutrients for further use by another generation of plants, fertility can be maintained more or less indefinitely —at least in theory. A very diverse population of organisms would be needed to ensure that the cycle of nutrients from the dead plants back to the living was complete. Alternatively, fertility could be maintained by one very complex organism, man, by continuously feeding the solution with new nutrients equivalent to those removed by the plants, but it can *not* be maintained in the complete absence of life other than the growing plant.

Soil fertility is thus a biological phenomenon. It cannot be created except by living organisms, and if a certain minimum quantity of life is not present it declines progressively. Fertility cannot usually be created by one type of organism, though, as we have seen, man can do so under exceptional circumstances. Under normal conditions the action of a very complex social organisation of living things is required, and the greater the productivity of the soil the more complex this organisation is. At one end of the scale we have the unproductive soils of the desert fringe or of the arctic tundra colonised by relatively few species of plants and small numbers of associated animals and micro-organisms, and at the other end of the scale some of the most productive soils in the world, created by highly organised human societies, with the help of the relatively few selected animal and vegetable species which serve them. The soil fertility which human societies create and which is the principal subject of this book is of the same kind and is the result of the same kind of biological forces as the fertility of a forest soil. The chief difference is that in one case the dominant organism in the living community is an animal and in the other a tree.

Human societies affect the properties of soil in many ways, but by far the most important is by the practice of agriculture. The object of permanent agriculture is to produce a limited number of specified plants which are required for the sustenance of men and domesticated animals. This involves the maintenance of the soil in a condition in which it will continue to produce agricultural crops for an indefinite period. None of the major arable crops of the world grows wild—that is to say, the habitat (the soil) has to be changed in some degree, if only by removing the existing vegetation or by scratching the soil with a stick, before such a crop will grow. If crops are grown on a large scale and to produce economic yields in a highly developed society considerable changes in the soil must usually be made. Almost any soil can be made to produce agricultural crops for a time, but no ' natural ' soil will produce them indefinitely. The basis of the numerous systems of shifting cultivation and other systems of temporary agriculture is the using-up of existing agricultural producing capacity without any attempt being made to perpetuate it. When the soil can no longer produce agricultural crops it is abandoned, and if it is not too far gone Nature will restore its pristine fertility in due time.

A soil capable of supporting permanent agriculture must have certain biological properties, imparted to it as a result of man's management and analogous to those different properties which enable a forest soil, for example, to support a forest indefinitely, or a prairie soil a prairie. In temperate regions certainly, and in tropical regions probably (though few examples of permanent tropical agriculture are known), the properties common to all permanent agricultural soils depend on the universal predominance in agriculture of graminaceous plants (cereals and grasses). Permanent agricultural soils, of course, vary widely in their general properties, depending on the environments in which they were originally found, but they must all acquire through the management they receive those special properties which enable them to produce human and animal foodstuffs century after century without depleting the soil's fertility. We have already said that fertility is the result of the action of living things on the soil, and in cultivated soils the principal agent in producing fertility is not, as in uncultivated soils, plants, but man, who by cultivating, manuring and selecting his crops can perform this primordial function much more effectively than plants can. When man does not create soil fertility—and being usually a member of a complex economic society he will only do so if it pays him—agriculture becomes merely a temporary expedient for rapidly exploiting the fertility accumulated during the formation of the soil, and can have no permanent basis. Thus, while all soil fertility is biological, agricultural soil fertility is also an economic phenomenon. In different climates different systems of management are required to develop the specific properties common to all soils fit for permanent agriculture.

Let us now follow the course of soil formation from its very beginning, which would be long before man takes a hand in it, because man likes to start his operations on soils already made fertile by previous forms of life—i.e. he is a late-comer among the living things which succeed each other in regular sequence in the process of soil formation. We shall take as an illustration a mineral soil because most agricultural soils are mineral soils, as distinct from peats (organic soils) or water soils.

A necessary precursor of mineral-soil formation is the weathering of rock exposed on the earth's surface. Weathering is (a) physical, whereby the rock is broken into pieces of different sizes down to microscopical by stresses produced by changes of temperature, freezing of water, friction with other pieces, etc. ; and (*b*) chemical whereby the rock is chemically attacked by slightly acid water containing carbon dioxide dissolved from the atmosphere. Physical weathering ultimately breaks the rock down to a state in which its particles can provide support for plant roots, and chemical weathering produces small amounts of two of the three major plant nutrients, phosphorus and potassium, in a soluble form in which plants can use them. Mineral rocks do not contain the third nutrient element, nitrogen, that plants must have, and it is still a bit of a mystery how all the nitrogen now contained in soils got there. But the first nitrogen probably came from the atmosphere, where it was combined with oxygen under the influence of electrical discharges and brought to the soil dissolved in rain water in the form of nitric acid, which is one of the forms in which plants assimilate nitrogen.

If the weathering rock is on a slope the fine particles produced by weathering will be washed down the slope by rain water until they reach the bottom of the slope or encounter some obstacle which prevents further movement. There they collect until a sufficient mass has accumulated for plants to root in. In time seeds will fall on the mass of rock particles, they will germinate and grow, using up all the small amounts of plant food produced by weathering. The plants will grow by absorbing energy from the sun, making, by the process of photosynthesis, the organic matter of which their leaves, stems and roots are composed. The plants get nitrogen, mineral salts and water from the soil, and carbon and oxygen from the air. When they die the dead residues fall on to the soil (as it may now be called), where they provide organic food for bacteria, fungi and other micro-organisms which decompose the plant residues, setting free the nigrogen and mineral plant food taken up by the living plants. The soil now contains all the plant food it originally had plus a little more produced meanwhile by the further weathering of the rock particles. It can therefore support a few more plants,

which in course of time will die and return to the soil all that
they took from it together with the organic matter they have
synthesised during growth. As this process of growth and decay
is repeated time after time, the plant-producing capacity of the
soil—its fertility—increases. At the same time the soil becomes
a storehouse of the solar energy absorbed by the plants during
photosynthesis, and this energy is utilised by an enormous
population of micro-organisms to release the nutrients contained in the dead plants for the use of future generations.

Soil fertility is thus essentially a cycle of solar energy, in
which the energy is absorbed by green plants in the form of
light, and used to build up complex organic compounds in the
plant tissues, whence it is returned to the soil to be gradually
dissipated as heat during decomposition, leaving a residue
of mineral plant food which gives the next generation of plants
an increased capacity of absorbing solar energy. The process
goes on until a state of maximum fertility is attained, at which
the intake of solar energy by the plants growing in the soil
is equalled by its dissipation by other organisms. This state
corresponds approximately to the appearance of the ' climax '
plant association—that is, the most highly developed plant
association that can be produced under the prevailing climatic
and geological conditions. In Britain the climax vegetation
is mainly deciduous forest in the south and coniferous forest
in the north, with scattered patches which are either too dry
for forest (e.g. the chalk downs) or too wet (e.g. peat bogs).

This is a highly simplified account of soil formation. My
purpose in writing it has been to show the essential nature
of soil fertility as the result of a process in which energy in the
form of light and heat is changed into forms which can be
used for the production of living matter and can be stored in
the soil until it is wanted. Under natural conditions [1] the
energy is derived from the sun, and the initial transformation

---

[1] I use the phrase ' natural conditions ' in the popular sense, i.e. the
conditions prevailing in the absence of man, though I do not consider that
men or the conditions they create are in any way unnatural.

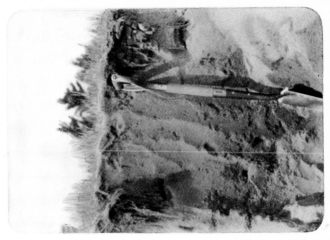

Plate 1b A brown forest soil
Photos : A. Muir

Plate 1a A podzol formed under
coniferous forest

whereby the energy is introduced into the soil cycle is made
by the green plant. An upper limit is set to the amount of
fertility that can be got into a soil by the amount of solar
energy that falls on to a unit area where the soil is formed.
When men create soils and soil fertility the process is fundamentally the same, but men are not confined to the sun as
a source of energy. They can take it from coal, oil, water
power or what you will, and they need not put it through
a green plant to turn it into fertility. They can make fertility
by dragging a plough across a field, or by using electric or
coal power to fix nitrogen from the air as immediately available plant food. In this way man can make more productive
soils than can plants alone, whose only source of energy for
creating soil fertility is the sunlight which falls on the ground
where they are growing.

Not all soils undergo the preliminary stage of weathering
of solid rock in the place where the soil is formed. Large areas
of Britain, for example, are covered by soils formed from parent
materials already crushed and partly weathered by glacial
action during the Ice Age, and transported often for great
distances before being deposited by the melting of the glaciers.
Soil parent material may likewise be transported and deposited
by wind or water, and may consist of a single kind of rock or
of many different kinds, and also of material that has already
been soil, but whatever the origin and nature of the parent
material, the soil is produced by the action on it of living
things. The different kinds of soils that can be produced by
this process are almost infinite in number and variety, as each
different combination of the possible variations in the three
main soil-forming influences—parent rock, life and climate—
will give rise to a different kind of soil.

Some time after the climax vegetation has appeared the
soil reaches a state of ' maturity ' at which the effects of
the vegetation, whatever its nature, on the character and
properties of the soil are fully displayed. Soil development, or
soil evolution as it is usually called, does not stop at maturity,

but further changes are slow in comparison with those which precede maturity. The pre-maturity stages are analogous to the stage of adolescence in the development of an animal, the maturity and post-maturity stages to the adult and senile stages. These may last for aeons in soil evolution. As a soil grows old it becomes worn out and less fertile. Its properties change, but its type characteristics remain the same, unless the biological environment undergoes a marked change.

From what has already been said it will be clear that there is a very close connection between the nature of a soil and the nature of the plant association it produces. The same close connection, of course, exists between the nature of a plant association and the nature of the soil it produces. Indeed, it is quite impossible to say which is the chicken and which the egg ! This connection is recognised in common parlance by such phrases as forest soil and prairie soil, or even agricultural soil and garden soil. A pine-forest soil is a soil made by and for a pine forest, and an agricultural soil is a soil made by and for agriculturists. We come across all sorts of pine-forest soils—heavy and light soils, thick and thin soils, rich and poor soils—but they all display certain common characteristics which enable us to include them in a single ' type.' Soils formed under some other coniferous forests (e.g. spruce) also belong to the same soil type. Similarly, different kinds of deciduous forest in temperate climates produce a well recognised type of soil which, though distinct from, is nevertheless related to, that of coniferous forests, both types being *forest* soils. Quite different are the grassland soils formed in climates too dry for tree growth, and different also are the many soil types formed under tropical conditions.

A characteristic common to all forest soils is that their surface layers tend to be leached of the more soluble chemical constituents, notably compounds of lime, potash and phosphate. A characteristic of natural-grassland soils is that their surface layers tend to be richer in these constituents than the subsoil. These and other soil characters are associated with the fact that forests occur in humid regions and grasslands in arid regions. A humid region is defined as one in which the annual amount of rainfall is greater than the annual evaporation from the soil, and an arid region as one in which evaporation exceeds rainfall. In a humid region, therefore, the movement of water through the soil, carrying dissolved substances with it, will be predominantly downwards, and in arid regions it will be, if anything, upwards. Forest soils consequently tend to be richest in plant food some way below the surface, where tree roots can get the nutrients, whereas grassland soils are richest near the surface where the main rooting system of herbage plants exists.

This difference between forest soils and grassland soils in the distribution of plant food to suit the needs of their respective plant associations is not produced entirely by the action of the plants, but also by the action of the climate. It affords a simple illustration of how closely interrelated climates, soils and plant associations are. When the climate changes both the soil type and the plant association change with it. It is the climate primarily that determines what sort of plant life will become established, and it is the plant life—and therefore also the climate—that determines what sort of soil will develop.

Each of the major climatic zones of the earth has associated with it a charactertistic type of plant association and a corresponding soil type. Going from north to south in the northern hemisphere we pass successively through the arctic, cold temperate, warm temperate (of which maritime and continental types may be distinguished), desert (a zone surrounding the 30th degree of latitude), sub-tropical and tropical zones. Here we shall consider the soil and vegetation types of the temperate zones only, with a few introductory remarks about arctic soils.

The arctic zone at the extreme northerly limit of plant growth has a vegetation consisting mainly of mosses and lichens, and a soil—the tundra soil—whose most obvious features are a layer of peat overlying a bluish-grey layer of mineral soil, the colour indicating poor aeration and drainage caused by the permanently frozen subsoil. This soil type, on account of the severe climate in which it occurs, is not used for agriculture to any extent.

The typical vegetation of the cold temperate zone is coniferous forest which stretches in a wide belt right across Europe, Asia and North America. The soil type associated with it is known as the *podzol*, and has a very striking appearance (Plate 1*a*). The name, like that of many other soil types, is Russian in origin, and means ' ashes underneath.' The podzol profile [1] consists of a layer of peat or ' raw humus ' [2] made up of the decomposing residues of vegetation overlying the most characteristic feature of a podzol—a bleached, white or grey layer which, it is said, Russian peasants used to believe was composed of ashes left by forest fires—hence the name podzol. Actually this layer is produced by the chemical action of organic acids—humic acids—derived from the raw humus on the mineral soil particles. In time the humic acids will dissolve from the soil practically everything except silica, which is left behind as a grey or white layer. The material dissolved by the humic acids, and consisting of the plant-food elements and also aluminium and iron compounds (the latter coloured red or brown), are carried downwards by percolating water and deposited in another, quite distinct, brown, red or black layer at the bottom of the bleached layer. The brown layer, which contains all the material removed from the bleached layer, is usually very compact and sometimes of rock-like hardness, in which case it is known as hardpan. It is quite rich in plant nutrients, but it also contains toxic aluminium and iron compounds which, together with the unfavourable physical condition, make it an unsuitable medium for plant roots to develop in. The main feeding zone in a podzol is at the boundary between the bleached and the brown layers.

---

[1] The term ' profile ' is used to describe what can be seen on a vertical section of a soil.

[2] Humus is fully explained in Chapter VII.

In Britain podzols can be found under coniferous forest, mainly in Scotland, which is in the podzol zone ; England is not. Nevertheless good examples of podzols can also be found in England and Wales on heathland, which forms on very acid rocks—for example, on the Yorkshire and Denbighshire moors, on the Bagshot sands in Berkshire and on Wareham Heath in Dorsetshire. The podzol is a very acid soil, and supports an acid-tolerant vegetation of coniferous forest or heath. Acidity in the soil-forming rock consequently favours the formation of a podzol.

Podzols are made by and for coniferous forests, for which they are quite fertile, chiefly because little else can grow on them. Like every other type of soil, they can be transformed into fertile agricultural soils, but podzols are only cultivated when no better soils are available. Much of the agricultural land of northern Scotland formerly had podzol soils, and was reclaimed with immense labour. The unimproved podzol areas of Britain are being increasingly utilised for the purpose for which they are economically best suited—the growing of pines and spruces.

South of the coniferous zone lies the deciduous-forest zone where rather different types of soil are found. They are ' brown forest soils ' in areas with a maritime climate, and ' grey forest soils ' in areas with a continental climate. Here we shall only consider the former which were originally the predominant soil type in Britain ; the latter do not occur in this country.

The appearance of a brown forest soil (Plate 1b) is not nearly so striking as that of a podzol. The surface layer is black or dark brown, and consists of well decomposed humus and mineral soil intimately mixed by the action of earthworms and insects. The mixture is known as ' mull,' and is very characteristic of brown forest soils, and quite different in appearance and properties from the peaty ' raw humus ' of podzols. The mull layer merges without any sharp change into the brown mineral-soil layer below it, and this again shades off gradually into the usually lighter colour of the parent rock.

The brown forest soil is regarded by some as an incipient podzol. The surface layer is mildly acid due to the washing-out of lime and other basic substances in the humid climate in which deciduous forest occurs, but there is not the far-reaching decomposition of minerals with removal of aluminium and iron compounds that is characteristic of the podzol process. If leaching of lime was progressive and acidity increased, a time would come when the mull was replaced by more acid raw humus, and then podzol formation would set in. But so long as the deciduous forest is there that does not happen, because the deciduous trees absorb large quantities of lime and other bases from the subsoil through their roots and return them to the surface soil when the leaves fall. Thus a balance is reached between the effect of rain in washing out bases (and plant nutrients) and the effect of the trees in returning them to the soil. The soil consequently remains quite rich and productive indefinitely so long as it is covered with forest. If the forest is removed, as happens of course when the soils are cultivated, the restorative action of the trees is lost and a deterioration of soil productivity sets in. This happened when British soils were first cleared and cultivated. The deterioration of the soils continued for centuries, and it was not until about four hundred years ago that means were found not only of preventing the deterioration but of raising soil fertility to a hitherto undreamed-of height.

Brown forest soils have many properties which make them suitable for agriculture, but they need constant care to keep them productive. They are the basis of western European agriculture, and have been used in western Europe ever since the people there learnt how to wield an axe effectively. English agriculture actually started in the Stone Age, before the invention of the iron axe, on the grass-covered chalk downs, but these were largely abandoned or given over to grazing when the brown forest soils became available.

South of the deciduous-forest zone we pass out of forest land into grass land. The natural grassland regions of the world occur mainly in the interiors of the continents, and are marked by their hot dry summers and cold wet winters. Trees do not thrive under these conditions. Many grassland soils are black in colour, and the most representative of them is known as *chernozem*, a Russian word which means black earth (Plate 2*a*). Chernozems are associated specifically with a steppe vegetation and a semi-arid climate. In many respects their properties are ideal for agriculture, and the reasons why, until quite recently, they have not been used are that the interiors of the continents where they occur were hardly accessible before the arrival of railways and motors, and that the climate, although excellent for spring-sown cereals, is too dry and harsh for the intensive culture of many other crops.

The true chernozem is a neutral soil, neither acid nor alkaline, and the most obvious feature of its profile is its uniform black colour, sometimes extending to a depth of several feet. It contains a fairly high proportion (say 10 per cent) of humus which is evenly distributed through the black layer and is very intimately mixed—probably chemically combined—with the mineral particles of the soil. This humus is derived partly from the overground parts of the grass vegetation, but mainly from its roots which penetrate the soil in all directions and are continuously dying off and being replaced. The grass root system is the most active agent in the formation of a chernozem, and imparts to the soil certain physical properties which are of fundamental significance to agriculturists in every country and clime.

These physical properties are expressed in what is known as the 'structure' of a chernozem. Soil structure can be defined rather crudely as the size and shape of the lumps into which a soil breaks up when it is thrown against a wall. A chernozem breaks up into hard crumbs of various sizes averaging, say, the size of a pea, that give the soil almost perfect physical properties for the growth of the great majority of agricultural crops. The crumbs are formed partly by physico-chemical reaction between the humus and the mineral particles of the soil, and partly by the mechanical action of densely growing grass roots which compress the soil into crumbs and subsequently, when the roots die and decompose, leave the crumbs separated from their neighbours with air spaces in between. The crumbs themselves are porous and have a high water-holding capacity, so that the soil is at once well drained and moisture-retentive—in other words, it always, except in conditions of extreme wetness or dryness, contains sufficient air and sufficient water to support vigorous plant growth. These favourable physical conditions also promote great microbiological activity during the growing season and rapid decomposition of humus with release of the plant food contained therein, so that plants are assured of everything they need—air, water and food.

Other soils have different structures. A podzol is either ' structureless,' with all its particles separate and unaggregated, or it has a laminar structure. This structure or lack of structure gives much less favourable conditions for plant growth than the crumb structure of a chernozem, but pine trees and their associated flora can tolerate them better than can grasses and most other types of plant, hence a podzol structure helps a pine forest preferentially to other, possibly competitive, plant associations. A deciduous forest makes greater demands than a pine forest on the water and nutrient supplies of the soil, and needs and makes a better structure for itself. The brown forest soil has a rather weakly developed crumb structure that breaks down easily when the forest is cleared and the soil is exposed directly to the sun and rain.

The reason why chernozems, of which the very fertile black earths of the Ukraine are well-known examples, are so pre-eminently suited for agriculture is that they are soils made by and for grasses, and all over the world grasses are by far the most important and most numerous of human-food plants —annuals (cereals) for direct consumption, and perennials for

Plate 2a
A chernozem

Photos : A. Muir

Plate 2b   An alkali soil (solonets), with salt-incrusted
'columns' exposed in the subsoil

Plate 3a
A gley soil

Photos : A. Muir

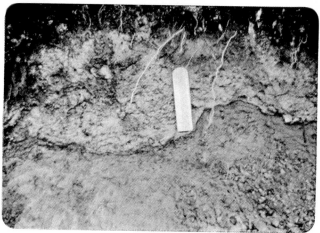

Plate 3b   A heath podzol with thin hardpan

consumption through livestock. A chernozem can be culti-vated for several years, and in both the wet and the dry state, without its structure deteriorating. Ever since permanent agriculture started, and wherever it has been established, the unconscious aim of the farmer has been to produce a soil with a stable crumb structure that would stand up to the climate. To achieve this aim different measures are required in different climates, and the art of agriculture has been to discover the appropriate measures and to incorporate them into an agricultural system that will work, economically and socially. This art is now becoming a science, for we know what is wanted, the formerly unconscious aim has become a conscious one, and to a large extent we can tell in advance what cultural, manuring and cropping practices will promote or prevent crumb-structure formation. But we cannot yet control the economic and social environment in which agri-culture operates, and though we may know what is needed to produce a soil fit for permanent agriculture in any region we may not be able to do it.

There are many other types of soil structure besides the crumb and laminar types already mentioned, but the crumb structure is, in the words of an eminent Russian scientist, A. N. Sokolovsky, ' the only one of all types of structure known that possesses agricultural value.' Soil structure is a very complex phenomenon, and is the resultant of all the factors —geological, chemical, physical and biological—active in soil formation. But the determining factors in both soil and soil-structure formation are biological. In the formation of a crumb structure one biological agent—a grass cover—is pre-eminent, indeed, might almost be said to be unique. No other way except by growing grass is known of producing a crumb structure that will be stable under cultivation, hence the increasing significance attached to grass as an agricultural crop in every country. Man is another biological soil-forming agent capable of producing a crumb structure, but he has not yet succeeded in doing so without the help of grass. How he has set about his job as a soil-forming agent in Britain, and how he has transformed the brown forest soils and podzols that formerly covered the country into some of the most fertile soils ever made is the theme of this book.

It must not be thought that every agricultural operation on the soil has for its object the creation of a good structure, but a good soil structure is the basis of permanent agriculture, and if a soil has not the proper structure it will deteriorate until agriculture becomes impossible. Breakdown of soil structure has been the cause of much of the soil erosion that has laid waste vast areas that have been wrongly farmed. Most farming operations, however, are much more concerned with getting a living for the farmer than with producing a soil structure. Growing sugar-beet does not promote structure formation, but that is no reason for not growing it ; though it will grow best in well structured soil, and that is a reason for maintaining a good soil structure. There are certain fundamental principles, which are identical with the recognised principles of good husbandry, that must be observed if a crumb structure is to be created and preserved under agriculture, for a soil with a stable crumb structure is synonymous with a soil in a state of high fertility.

CHAPTER II

THE ARCHITECTURE OF SOIL

SOIL is the habitat or living quarters of a vast population of animal and vegetable organisms ranging in size from a gigantic tree to a microscopic bacterium. Every organism has to find or make suitable living quarters for itself or perish. The first-comers find simple and crude quarters in the spaces between the loose rock particles of which a primitive soil is composed. As the soil population enlarges and becomes organised it produces its own building materials from the bodies of its members, with the help of which each organism creates a suitable home —a pore space—in which it can eat and drink and breathe. The soil thus acquires an extraordinarily complicated physical structure comprising a framework of solid material enclosing an almost infinite variety of pores and channels. The action of plant roots in pushing through the soil is an obvious example of how organisms make room for themselves to live in, but all other organisms also play parts in forming a soil structure. Different soil populations will produce different kinds of structure ; the matted roots of grasses will tend to produce a crumbly structure in the surface soil, and the larger roots of trees a cloddy structure in the subsoil.

The difference between the structure of a grassland soil and a forest soil is analogous to the difference between the structure of an industrial city and a garden city where the different ways of living of the inhabitants lead to differences in the architecture, size and spacing of the buildings and in means of transport and trade. These are the things a sociologist would study rather than the bricks and mortar of which the buildings were made. The student of the soil pays more attention to the architecture of soil lumps and crumbs, and particularly to the spaces between the crumbs in which air, water and plant nutrients occur, than to the ultimate particles of rock from which the soil buildings are constructed. All the actions and reactions in a soil go on in the pore spaces between and within the soil crumbs and on the solid surfaces which bound the pores. The fertility of a soil is largely controlled by the architecture of the pore space, which in turn is determined by the size, shape and packing of the lumps, crumbs and particles of soil as they occur in the field.

THE SIZE OF SOIL PARTICLES

The nature of the bricks—the ultimate rock particles—and mortar—the very smallest, colloidal particles—of which a soil is composed and with which the population builds its homes is, however, not without significance, especially in agricultural soils in which the farmer uses the bricks and mortar to alter the structure every time he cultivates. One of the purposes of cultivation is to produce a structure in which seeds will germinate and grow easily, and the cultivator needs to know something about the physical nature of his building materials.

Soil can consist of mineral particles of any size from big stones, which are chemically quite inert and physically nearly so, to gravel, sand, silt and clay, in descending order of magnitude. The smaller the soil particles are the more active they are chemically and physically. This is because nearly all the chemical and physical activity of a soil takes place on, and depends on the magnitude of, the surface of the particles. If we assume for simplicity that soil particles are all cubes (it makes no difference to the argument that they never are), then a cubic inch of soil consisting of one cube only will have a total surface of its six faces of 6 square inches. If the cube is broken down into cubes with sides of $\frac{1}{10}$ inch, 1,000 such cubes will occupy a cubic inch, each cube having an area of $\frac{6}{100}$ square inch. The total surface in a cubic inch will then be 60 square inches, and each time we divide the length of the cube by 10 we increase the total surface contained in a given volume 10 times. A cubic inch of soil consisting of cubes of one ten-thousandth of an inch in length would expose a total surface of 60,000 square inches or nearly $\frac{1}{100}$ acre of soil.

The size of particles can be roughly measured by the speed at which they fall when suspended in water ; the smaller they are the more slowly they fall. It is possible to calculate, from the rate at which particles settle from a water suspension, what their diameters would be if they were spherical. In this way a complete ' mechanical analysis ' of a soil can be made, the mechanical composition being expressed by the percentage of each of several classes or fractions of soil particles present. In making a mechanical analysis the structural aggregates composed of numerous particles stuck together are dispersed by physical or chemical means so that each particle sinks separately through the water suspension. The size limits of the different classes or fractions of soil particles commonly distinguished in Britain are :

Gravel	greater than 2	mm. diameter	
Coarse sand	2–0·2	,,	,,
Fine sand	0·2–0·02	,,	,,
Silt	0·02–0·002	,,	,,
Clay	less than 0·002	,,	,,

The lower limit of the clay fraction is zero, clay particles can be so small that they cannot be seen under the most powerful microscope, but ultramicroscopic clay is seldom present in sufficient quantity to affect the agricultural properties of a soil. Humus particles are, however, ultra-microscopic, and it is their minute size which gives them such great chemical and physical activity.

All particles, whether of clay or humus, smaller than 0·002 mm. (or about one ten-thousandth of an inch) diameter are known collectively as the ' colloidal fraction,' and the particles themselves are known as ' colloids.' Colloids possess, by virtue of their finely divided state, great surface activity and have profound effects on the behaviour of soils containing much of them. They are the principal cementing agents in aggregating the larger particles into crumbs. A soil containing no colloids—for example, a pure sand devoid of humus—would be unable to retain in its mass either water or dissolved plant nutrients, and would consequently be unproductive if treated by ordinary agricultural methods.

Soils are sometimes found that can be classed as pure or nearly pure sand, silt or clay soils—that is to say, a very large proportion of all the particles falls within the size limits of the corresponding textural fraction. But most soils contain appreciable quantities of several or all fractions. When the sand and clay fractions are nicely balanced the soil is called a loam, and for most agricultural purposes a loam, which combines the virtues of a light soil with those of a heavy soil without their corresponding disadvantages, is preferred. On the other hand a pure silt soil, composed of particles all intermediate in size between sand and clay, has most of the disadvantages and none of the advantages of a mixture of sand and clay. Silt is too coarse to enter into combination with humus to produce the complexes which form a crumb structure, and it is too fine to be porous. Consequently it is very difficult to prevent a soil rich in silt from compacting so that neither air nor water can get into it.

Soil Texture

What is commonly known as the 'texture' of a soil—whether it is light or heavy, sandy, loamy or clayey—is determined primarily by the relative proportions of different-sized particles in its make-up, but also by the state of aggregation or structure of the soil. The terms 'light' and 'heavy' refer to the amount of power required to draw a cultivating implement through the soil. A light soil has a large proportion of coarse particles, and a heavy soil a large proportion of small particles, the lightness being caused by the relatively large air spaces existing in a mass of large particles compared with the greater compaction obtaining in a mass of small particles. A clay soil, however, under comparable conditions of deposition, will contain a greater percentage of pores per unit volume than will a sand soil ; the pores will be smaller, but much more numerous. The amount of pore space—that is, the amount of space not occupied by solid matter—is often within the range of 35 to 50 per cent of the volume of the soil. Presence of organic matter (humus) tends to increase the porosity. Attempts have been made to define precisely, in terms of the proportions of the different size fractions, what is meant by a sandy soil, a sandy loam, a clay loam, a clay soil and so on, but in this case precise definitions are not much use. The agriculturally important physical properties are determined more by soil structure than by particle-size distribution, and can be assessed, after a little practice, more satisfactorily by rubbing a little moist soil between the fingers. This is sometimes referred to as the ' field texture.'

As a general guide it may be said that a soil would be described as sandy if it contained 60 per cent or more of the sand fraction, as a loam if it contained appreciable sand and not more than 30 per cent of clay, and as a clayey soil if it contained over 30 per cent of clay and less than 50 per cent of sand. These limits are very approximate, but they indicate the relative importance of the clay fraction in determining the physical properties of a soil. It does not need nearly as much clay as sand to become the dominant fraction.

A light sandy soil suffers from the drawbacks of too great permeability, causing excessive drainage and a tendency to dry out readily, and too low water-storing capacity, which again causes crops to suffer in periods of drought. It is liable to be poor in mineral plant nutrients. The liberal aeration characteristic of sandy soils results in a rapid oxidation of humus, that is further promoted by the ease with which such soils warm up. Sandy soils are often described as ' hungry '—for organic manure and the plant nutrients contained therein. Heavy clayey soils, on the other hand, are poorly permeable or impermeable and suffer from deficient drainage. Chemically, they are usually quite rich in plant nutrients which, however, owing to bad aeration may not be in a condition available to plants. For the same reason they do not respond to applications of manure and fertilisers as readily as do sandy soils. They are liable to waterlogging, the clay particles swell when wetted, and the whole soil may be reduced to a structureless slush. Heavy soils are cold soils because, especially at the end of the winter when warmth is needed to start plant growth, they are full of water which takes more heating up than any other substance known.

Crops which like or will tolerate sandy soils include oats, rye, lupins, early vegetables and potatoes.

Crops which do well on heavy soils include wheat, oats, barley, beans, mangolds, grass.

The ideal degree of both permeability and water-holding capacity is most likely to be found in a mixture of sand and clay (a loam) with a considerably higher proportion of sand than of clay. Other things being equal, such a mixture will contain the most suitable proportion of coarse pores through which water can drain to fine pores in which water is held by capillary forces. It aggregates readily and provides a physical stratum in which almost every plant can root and grow freely.

The ultimate texture (particle-size distribution) of the subsoil is perhaps more important to the agriculturist than is that of the topsoil, because the actual field properties of the topsoil are the result not only of particle-size distribution, but also of the development of a soil structure by the aggregation of the soil particles into crumbs and lumps which profoundly modify the texture. A clay soil made permeable by having a good structure may overlie a subsoil derived from the same geological material and weathered to the same degree, but quite impervious to water, and causing the whole soil to suffer from waterlogging. Equally, a sandy or gravelly subsoil may cause excessive drainage.

In the majority of cases the soil and the subsoil consist of similar material, but not always. It is quite possible for a

Plate 4   Some types of soil structure
(*above*) platy ;   (*below*) blocky
Photos :   U.S. Department of Agriculture

Plate 5   Some types of soil structure
(*above*) prismatic ;   (*below*) crumb
Photos :   U.S. Department of Agriculture

sandy subsoil to underlie a heavy topsoil, or *vice versa*. An excessively heavy or an excessively light topsoil will then acquire some of the characteristics of a loam in the sense that the poor drainage of the former will be improved by a permeable subsoil, and excessive drainage in the latter will be prevented by an impermeable subsoil. In forests, especially, a heavy, water-retaining subsoil is often a great advantage. Wilde (1950) [1] found that most soils with compacted subsoils supported nearly twice the volume of trees carried by soils with pervious subsoils.

The Minerals in Soil

The mineral composition of the different size fractions varies considerably. In the coarser fractions, down to silt, by far the commonest mineral is quartz or silica (SiO_2), partly because it is the most resistant to decomposition. Other common minerals occurring in the sand and silt fractions are the feldspars, which are aluminosilicates of potash, soda and calcium, and the micas which are rather differently constituted aluminosilicates of potash and magnesia. Quartz contains no plant nutrients, consequently light soils composed mainly of coarse or medium-sized particles tend to be ' hungry ' soils, requiring large amounts of manure or fertilisers. In any case, the nutrients contained in coarse particles are not readily available to plants. The coarser fractions consist largely of unweathered material, or, rather, of the only slightly altered residue which is left behind when the disintegrating and dissolving weathering agents have done their work.

The minerals found in the finely divided clay fraction are quite distinct from those found in the coarser fractions, and are the products of the chemical action of weathering agents and living organisms on the original rock particles. Several known minerals have been identified in soil clays, but they can all be put into one of three groups—the montmorillonite

[1] See Bibliography, p. 215

group, the mica group or the kaolinite group. They are crystalline aluminosilicates of more or less complex chemical structure, and their physico-chemical properties and those of the soil clays containing them depend on the group to which they belong. Montmorillonite-type minerals show the greatest, and kaolinite-type minerals the least, activity, and many of the properties associated in clays with soil fertility, such as the capacity to absorb exchangeable cations (p. 37), to form clay-humus complexes and to produce a crumb structure, are displayed most sharply by montmorillonite minerals which also give rise to the smallest crystals. Which type of mineral occurs or predominates in a soil clay depends on many factors of which the chief are the nature of the rock from which the clay is derived and the course of weathering. Montmorillonite is characteristic of an early stage of weathering, mica of an intermediate stage and kaolinite of an advanced stage.

Soil Structure

The texture gives little indication of a soil's fertility. Nevertheless, most people, with a little practice, can form an idea of whether a soil is good, bad or indifferent from superficial observation. They observe whether the soil is ' in good heart,' has a good ' tilth ' or appears ' ripe '—expressions which reflect a general impression that the soil is a favourable habitat for plants. The same general impression may be given equally by a sandy or clayey soil, and derives mainly from the observer's intuitive conclusion that plant roots can get from the soil the air and water they need for growth. What is being observed is the soil's structure. A crumby porous soil, in particular, gives an impression of fertility, and the impression is strengthened if the soil has a dark colour indicating the presence of ample humus. A compact clay or loose sand, on the other hand, gives an impression of low fertility.

It is impossible to define soil structure precisely, because structure is a combination of all the physical, chemical and biological properties of soil. We have already described it unprecisely as the size and shape of the lumps into which a mass of soil disintegrates. The texture—the size of the ultimate particles—has an obvious influence on structure, and particularly the amount and nature of the colloidal clay and humus which represent the mortar with which the larger particles are cemented together to form compound aggregates. The chemical properties, particularly the reaction and the nature of the exchangeable cations which can flocculate or disperse the colloids, causing the mortar to ' set ' or ' run,' also influence structure ; but the biological properties which represent the result of the activities of all soil-inhabiting organisms have in the majority of soils the greatest influence because the organisms make the structure for themselves. It is, however, unwise to try to separate physical, chemical and biological influences on structure formation, since the biology of soil is conditioned by its physics and chemistry.

The scientist must have some method of describing and measuring soil structure. In the most common method the amount of crumbs of different sizes that are not broken down when shaken gently in water are measured. The water stability of crumbs has special agricultural significance because crumbs which are dispersed by rain will have little permanence or value, but the measurement of water-stable crumbs gives only the crudest picture of soil structure. What is really needed is a picture of the complete system of the pores and capillaries in which the biological activities of the soil take place. At present no practicable method of making such a picture has been devised, though Kubiena (1938) has developed a system of microscopic study of undisturbed soil that holds out great promise, but is not suitable for routine work and gives almost too detailed a picture for everyday purposes.

Different types of soil have different types of structure ; indeed, the structure is the most typical characteristic of a soil, embracing as it does most of the other soil properties. There is the almost structureless or ' platy ' condition of a

podzol surface soil, and the 'blocky' condition of the compacted subsoil (Plate 4*a*, *b*), caused by acidity and the
washing-out of the soil colloids. A complete lack of aggregation (' single-grain ' structure) is found in salt-containing soils
where the salts change the physical state of the colloids so
that they cannot bind the particles together. Alkaline soils
display a very characteristic cloddy ' columnar ' structure,
illustrated in the subsoil in Plate 2*b*, and caused by the drying-
out of dispersed clay which has been deflocculated by the
action of alkalies. A somewhat similar, but more angular
' prismatic ' structure (Plate 5*a*) often occurs in soils of dry
regions, and is caused, at least partly, by excessive drying-out
of the clay particles.

The structure that the majority of soil-inhabiting organisms
—animal, vegetable and microbial—thrive best in is the
' crumb' or ' granular ' structure (Plate 4*b*) described briefly
on pp. 13–14, and more fully in Chapter X. They cannot,
however, always make the ideal structure because the nature
of the bricks and mortar—the physical and chemical properties of the soil particles—prevents them. A crumb structure,
in which the rock particles are aggregated into small hard
crumbs allowing both the retention in the soil and free per-
colation through it of air and water, is most readily formed
when the soil contains sand and clay in suitable proportions
(a loam), humus to bind the particles together, and sufficient
lime to flocculate the clay and humus colloids. Thus in very
acid soils containing little lime and in alkaline soils containing
soda the colloids are dispersed, the mortar between the bricks
does not ' set ' and soil crumbs are not produced. Only a few
kinds of plant and no common agricultural crop can thrive
under these conditions. If a soil consists entirely of sand or
silt there may be insufficient colloid mortar to produce a
structure, though plants can themselves produce some mortar
in the form of humus ; and if a soil consists mainly of clay
there may be too much mortar and too few bricks for crumb
formation, though here again humus can work wonders.

Plants help to make a structure by providing humus, which is a more efficient binding agent than clay, and by penetrating the soil with their roots, pressing the particles together and breaking them up into crumbs. Animals and insects also provide humus when they die, but their chief structure-forming function is to ingest dead plant residues, to carry them into the soil and to excrete them as humus. Earthworms are pre-eminent in this respect (Chapter V). Micro-organisms of all kinds take part in the decomposition and humification of plant and animal residues. Each member of the immense and diverse soil population, in collaboration or competition with every other member, uses the inorganic and organic materials available to make the best living quarters for itself. The result is a soil structure which provides tolerable living conditions for all, but seldom ideal conditions for any.

A crumb structure is essential for intensive agriculture. Only in a well structured soil can crops find the conditions that will allow bountiful yields to be obtained. The first objective in soil management is to produce a crumb structure without which full benefit will not be derived from any other virtue the soil may possess. The farmer adopts a specific system of cultivation, manuring and crop rotation that will produce the best structure that can be made from the materials of which his soil is composed. He is a much more powerful structure-former than the natural vegetation. He can change the texture of his soil ; he can add lime and other chemicals to flocculate or disperse the colloid ' mortar,' and as much or little humus as he needs ; and he can break up or compress the soil as much as he likes (though not always with the desired result). But his most trustworthy agent is grass, the matted root system of which can nearly always be relied on to produce a crumb structure. It is for this reason that the inclusion of perennial grass in the rotation (ley farming) is now being strongly advocated in many countries. All the other mechanical and chemical contrivances at the farmer's command cannot equal grass as a structure-former.

THE SOIL PROFILE

A noticeable feature about the architecture of soil is the differentiation of the vertical section or 'profile' into more or less distinct layers or 'horizons' which are often quite independent of any geological stratification that may be present. In a humid country like Britain the differentiation of soil horizons is caused primarily by the washing-out from the surface layers of soluble constituents and sometimes also of finely divided colloidal clay particles, and their re-deposition lower down in the soil. In this way an 'eluvial' or 'A' horizon is formed on the surface, and an 'illuvial' or 'B' horizon underneath. The horizons are very well shown in the podzol profile (Plate 1a). In the podzol there has been washing-out not only of water-soluble compounds, but also of water-insoluble compounds which have been dissolved from the rock particles by humic matter percolating from the peaty layer on the surface. The different chemical composition of the two horizons produces a different structure in each, the removal of colloidal matter from the 'A' horizon producing a loose, almost structureless condition, and the precipitation of colloids in the 'B' horizon producing a hard compact structure. Differences in chemical composition may also result in differences in the floral and faunal populations of horizons, that will give rise to structural differences.

In all soils the humus modifies the leaching effect of percolating rainwater, sometimes intensifying it, as in the podzol, sometimes retarding it. Thus soil horizons are the result of the prolonged influence of rainfall (or, more accurately, climate) and of the soil-inhabiting organisms, animal and vegetable, which produce humus and react upon the soil structure. Horizon differentiation will vary according to the climate and the nature of the soil organisms and, as we have seen in Chapter I, different types of soil, distinguished by their different architecture, are found under different types of climate and vegetation.

Below the ' B ' horizon is, as one would expect, the ' C ' horizon, which is not really soil at all and consists of disintegrated, partly weathered rock particles which contain few or no organisms, but will become the ' B ' horizon as the profile deepens in course of time or if the surface layers are removed by erosion. The ' C ' horizon usually begins where humus penetration ceases. The ' D ' horizon, below the ' C,' consists of unweathered rock, and will gradually be merged into the ' C ' as weathering proceeds.

Horizon differentiation is most clearly seen in the podzol, where the colours of the ' A,' ' B ' and ' C ' horizons are quite distinct and the boundaries between them are sharp. In many other types of soil, including the common British brown forest soil (Plate 1b), the horizons merge gradually into one another and sometimes can only be distinguished by chemical analysis. In agricultural soils the original horizon is usually disturbed by cultivation, which may also take in part of the ' B ' horizon, but the nature of the latter, especially if it is compacted as in a podzol, may have an important bearing on the agricultural properties of the soil. Continued cultivation forms, down to the depth of cultivation, its own ' A ' horizon, which is homogeneous, and, if the soil is manured, richer in plant nutrients than the underlying subsoil. Continued ploughing to constant depth tends to produce a ' plough pan ' of compacted soil which forms a sharp boundary between the arable horizon and the subsoil. The cultivated horizon is, of course, the horizon in which man's influence is predominant.

On the face of a soil profile is written the history of that particular bit of soil. It is not an easy history to read, because although every event leaves its record on the profile, each particle of soil registers countless successive events which may modify the record of preceding events. From the nature of the strata and of the minerals composing the soil we can tell its geological history and the type of weathering the strata have undergone. From the nature and number of the horizons we can deduce something of the living things that have inhabited the soil in the past as well as in the present ; if, for example, we find a podzol under grass vegetation we can assume that the soil was originally formed under forest. Profile interpretation is a fascinating art and can be most informative, but for practical purposes the main points to notice in a soil profile are, first, the differentiation of the soil into horizons, and secondly, the colour, texture and structure of each horizon. The colour provides information about the amount and quality of the humus, as well as telling us something about the condition and properties of the soil minerals ; the texture tells us about the ease of working of the soil, its permeability and drainage capacity, and perhaps what crops not to grow ; the structure tells us whether the soil is already a fit habitat for exacting agricultural crops, and, if not, what kind of treatment will improve it. These are rough and ready indications, but in the hands of an expert interpreter can provide very valuable information. For more precise information—for example, what and how much fertilizer to apply, and why some plants and animals will not thrive on a certain soil—recourse must be had to a laboratory examination.

CHAPTER III

THE CHEMISTRY OF SOIL

WHEN you are faced with the necessity of having to make sense out of the chemistry of the soil, it is helpful to know at the start that the parts of that science that are useful to agriculture are in the main quite simple to understand, and the parts that are not useful are more difficult. A great deal of soil chemistry, including many recent advances, has little or no immediate application to agriculture, though it probably will have in the future. In a book published about thirty years ago Bertrand Russell, speaking of science in general, said that the remarkable thing was not that we knew so much, but that, knowing so little, we could do so much. The remarkable thing about present-day soil chemistry is not that we know so little, but that, knowing so much, we can do so little with it. By analysing a soil we can—not directly from the analytical figures, but by a little quite legitimate juggling with the results—forecast fairly accurately how crops will respond to fertilisers, or which plant nutrients are in short supply in the soil. We can get the same information more reliably by growing crops on the soil for several years and observing directly how they respond to fertilisers, but this is a much more tedious and time-consuming procedure. Farmers could save themselves considerable sums on their fertiliser bills by having their soils analysed by the National Agricultural Advisory Service and acting on the recommendations sent with the analysis, even though the recommendations are mostly quite empirical.

The chemistry of much of the soil has little interest for the practical agriculturist. It is only the chemistry of the particles smaller than about a ten-thousandth of an inch in diameter that need concern us here. These include particles of molecular dimensions, such as matter in solution, and of near-molecular dimensions, such as humus and colloidal clay particles. Humus deserves a chapter to itself, and will be referred to only incidentally here.

The small inorganic particles are somewhat different chemically from the larger particles. They are composed essentially of silicates of aluminium, iron, magnesium, potassium, titanium and many other elements. The great majority of soils contain ample supplies of all the elements, of which there are at least fifteen, required for plant growth for at least a hundred years, and as the elements are used by living organisms over and over again, soil exhaustion rarely occurs except under agriculture in which abnormally large quantities of vegetation are harvested and removed from the land every year. The proportion of the total supply of plant-nutrient elements in the soil involved in the growth of crops or vegetation is, however, very small. The great bulk of insoluble silicate material comprising most mineral soils takes no part at all in plant nutrition. The active ingredients are the substances dissolved in the soil solution or retained by physico-chemical adsorption on the surface of colloidal clay and humus particles.

PLANT NUTRIENTS

These active ingredients take part in an almost completely closed biological cycle of (1) absorption of nutrient elements from the soil by plant roots and translocation of the elements into various parts of the plant, (2) return of the elements to the soil when the plant drops its leaves or dies, (3) decomposition of the plant residues by living organisms and the ultimate release of the nutrient elements in forms available for feeding another generation of plants.

So long as the biological cycle remains closed plant nutrition can look after itself indefinitely, but if the cycle is broken, as it has to be in most kinds of agriculture, plant nutrients are lost from the soil and compensating measures have to be taken to replace the losses. There is nothing we can do to tap the vast reserves of nutrients locked up in the silicate material comprising the bulk of soil that will only be released by slow weathering. The chief way in which soil chemistry can help the farmer is by assessing the magnitude of the annual or periodical losses from the biological cycle and finding means of replacing them. Replacement is almost entirely a matter of applying fertilisers and manures.

The effects of breaking the biological cycle are well illustrated by a case which happened in India a few years ago. Some native milch cows were doing very well on a local pasture, but were very poor milk yielders, and an enterprising official decided that they should be replaced by more productive animals. Within a few months of the change being made the new cows began to show symptoms of bone malformation that are associated with a deficiency of phosphorus in the diet. What had happened was that a large amount of phosphorus had been removed from the soil-plant system in the milk of the high-yielding cows—phosphorus that with the low-yielding cows would have been returned to the soil in urine. The soil was unable to replace the loss of phosphorus from its unweathered phosphate reserves, the phosphorus content of the pasture fell, the cows could not get enough phosphorus to satisfy all their needs, and, as milk has priority over bone for the nutrients in the blood stream, the bone structure suffered. The deficiency could be remedied by applying phosphate fertilisers.

A hundred years ago it was thought that a chemical soil analysis would give all the information required to prescribe the fertiliser treatment that would give bumper crops on any soil. All that would be needed was to find out, from the chemical composition of a crop, how much of each plant nutrient is required for full growth, and to supply these amounts in fertilisers, taking into account the quantities of nutrients already in the soil, as shown by chemical analysis. But it was found that the total quantities of plant nutrients—nitrogen

(N), phosphorus (P) and potassium (K)—in the soil bore no relation to the quantities that plants could absorb, and that plants used quite small and very variable proportions of the nutrients applied in fertilisers. Some of the many different nitrogen, phosphorus and potassium compounds already present in soil can be absorbed by plants, some cannot, and some are in process of becoming absorbable (or ' available ') or non-absorbable (or ' unavailable '). It is safe to assume that the nutrients in completely insoluble compounds, like the phosphatic mineral apatite or the nitrogenous parts of humus, are unavailable to plants, and that soluble compounds, like the phosphates in freshly applied superphosphate, or nitrates, are available. Actually the distinction between available and unavailable nutrients is not clear-cut. The soil chemist, when he gives the farmer an analysis showing the amount of ' available phosphate ' in parts per million of soil, is giving a figure which has no real meaning in itself, but is useful for comparison with figures for other soils of known behaviour. Thus a soil analysing at 80 parts per million of available phosphate might respond to treatment with superphosphate whilst a soil with 200 p.p.m. would not.

What the chemist does, in fact, is to extract the soil with some standard reagent like dilute acetic or citric acid, and find out how much of the plant nutrient is dissolved by it.

Dilute acids are used because it was once thought that plants obtained mineral nutrients from the soil by excreting dilute acids which dissolved the soil minerals. It is unlikely that this actually happens to any extent, but what matters is that the composition of the extract should reflect the relative likelihoods that different soils will respond to fertiliser applications. The phosphate in the extract is known as the ' available ' phosphate, but its only connection with availability to plants is that soils showing a low available-phosphate content usually respond to phosphatic fertilisers, and those showing a high available-phosphate content do not. The idea of nutrient availability is, however, very useful ; it expresses in one word everything that we do not know about plant nutrition, and that would be incomprehensible to all but a very few if we did know it.

Using these crude but practical methods of measuring the availability of soil nutrients, it is found that quite small proportions of the total nutrients are available at any moment. Thus an average amount of total phosphate (PO_4) in a soil may be 0·2–0·4 per cent or 2,000–4,000 parts per million, whereas the amount of available phosphate (soluble in dilute acetic acid) may be only 100–200 p.p.m. The proportion of available to total potash (K_2O) is of the same order, and of available nitrogen (N) much less.

The following figures, taken mainly from Leeper (1948), show the amounts of the three main nutrient elements, N, P and K, removed from the soil by wheat and milk in comparison with the total amounts present in soil. Figures are in pounds per acre and refer to the top six inches of soil.

	N	P	K
Total in soil	1,000–10,000	500	2,000–20,000
Removed in 25-bushel wheat crop	19	3·6	6·3
Removed in milk (cow to 2 acres) annually	18	2·3	3·7

These figures relate to Australian conditions. For British farming with its higher yields the figures for removal from the soil would be proportionately higher, but so also would be the figures for total nutrients, since the higher yields are obtained mainly by adding N, P and K to the soil. It will be seen that the quantities removed are almost negligible compared with the total in the soil, nevertheless it often only requires two or three crops to be grown without fertilisers to produce symptoms (falling yields) that the nutrient supply of the soil is being exhausted.

The effects of cropping and manuring on the nutrient content of the soil are shown by the following analyses of Broadbalk field, Rothamsted, in three different years. Broadbalk has carried wheat with different manurial treatments since 1843, thus the last analyses, in 1893, were made after the fiftieth successive wheat crop. The figures are the percentages of potash (K_2O) and phosphoric acid (P_2O_5) extracted from the topsoil by 1 per cent citric acid, and may be assumed to represent the amounts 'available' at the time of sampling.

TABLE I — PERCENTAGES OF K_2O AND P_2O_5 EXTRACTED FROM BROADBALK TOPSOIL BY 1 PER CENT CITRIC ACID

	Farmyard manure	No manure	N only	N+P	N+P+K
K_2O					
1865	·030	·0040	·0040	·0025	·0200
1881	·028	·0032	·0020	·0020	·0228
1893	·038	·0032	·0032	·0040	·0188
P_2O_5					
1865	·0355	·0094	·0106	·0259	·0261
1881	·0372	·0074	·0068	·0329	·0383
1893	·0560	·0078	·0074	·0405	·0434

The amounts of available nutrients closely follow the manuring, and show that considerable quantities of potash and phosphate accumulated in the plots where they were applied every year. The cumulative effects of farmyard manure were especially large. The effects of cropping in exhausting the nutrient content are shown by the third column (no-manure plot), and are quite small, but yields on this plot and therefore the amounts taken up by the crop were also small.

EXCHANGEABLE CATIONS, ADSORPTION

As mentioned in Chapter II, the chemically active part of the solid matter of soil comprises the very small colloidal particles (conventionally defined as those smaller than ·002 mm. in diameter), which by virtue of their smallness expose a large area of surface relative to their mass. Reactions between liquids and solids, and most soil reactions are such, take place at the surface of the solids. If the solid is rapidly dissolved by the liquid, as when sulphuric acid reacts with zinc, a fresh solid surface is continually being exposed, and the state of subdivision of the solid is not a vital factor in getting the reaction to go at all. But if the solid is only imperceptibly attacked by the liquid, as when rock is weathered by water, the state of subdivision of the solid becomes of prime significance. Thus, if we shake coarse rock particles with a solution of salt, wash the salt away and repeat the process several times, nothing perceptible will happen, but if the same rock is first ground to the colloidal dimensions of clay and given the same treatment, it will undergo a profound physical and chemical change which would affect the properties of any soil of which the clay was a component.

This reaction between salt and clay is a type of reaction known as base exchange or, more correctly, cation exchange, and is of the greatest importance in soil chemistry. The phenomenon of cation exchange in soils was discovered in 1850 by J. T. Way when he was seeking to discover how the recently introduced artificial fertilisers worked. Being very soluble in water they should mostly have been washed out of the soil before plants could absorb them, but they were not. Way found that when a solution of ammonium sulphate $(NH_4)_2SO_4$, was washed through soil, what came out at the other end was a solution of calcium sulphate, $CaSO_4$, the ammonium, NH_4, being retained by the soil and ' exchanged ' for calcium, Ca, already in the soil. Way showed that potassium (K) and phosphate (PO_4) were also retained by soil, but nitrate (NO_3) was not.

The explanation of this phenomenon subsequently put forward was that the soil colloids, which include not only the minute clay particles but also the still more finely dispersed and active humic matter, possess the property of removing, by adsorption on the colloid surface, the metallic component or cation of a salt from its solution, and substituting for it another cation already adsorbed on the colloid. Thus when ammonium sulphate is poured through soil the ammonium cation (NH_4) is ' exchanged ' for calcium ion already adsorbed on the soil particle, and the calcium combines with the remaining sulphate (SO_4) part of the ammonium sulphate to form calcium sulphate ($CaSO_4$) which appears in the drainage water. The best known use of cation exchange is in the water softener in which calcium cations which make water hard are exchanged for sodium cations absorbed on permutite. When all the sodium has been exchanged for calcium, a solution of sodium chloride (common salt) is poured through the water softener, the absorbed calcium on the permutite is exchanged for sodium, and the water softener is ready to function again. There are many industrial uses for cation exchange, for example, in extracting magnesium from sea water, but none is half so important as its function in agriculture, because it is the means whereby mineral plant nutrients are stored in the soil.

The mechanism of cation exchange is explained by assuming that the soil colloids are composed of negatively charged particles surrounded by a layer of positively charged cations which counteract the negative charges and leave the whole system electrically neutral. The negative charges are represented by acidic anions,[1] mainly of complex silicic and humic acids, and the positively charged by the cations calcium, magnesium, potassium, sodium, ammonium and hydrogen.

[1] In electrolysis, negatively charged anions move to the anode and positively charged cations to the cathode of the electrolytic cell.

A clay particle could be represented diagramatically as

The presence of absorbed hydrogen cations might make this clay rather acid, since exchangeable hydrogen is the main cause of soil acidity. To some extent the acidity would be neutralised by the other cations which are all basic, inducing alkalinity.

If a soil containing clay of the above composition is treated with ammonium sulphate the reaction represented below may take place :

ammonium being absorbed by the clay, and magnesium, sodium and potassium sulphates and sulphuric acid appearing in the drainage water. If a large amount of ammonium sulphate is used, all the cations on the clay particle can be exchanged for ammonium, but calcium and hydrogen are the most firmly held of the cations and go last, preceded by magnesium, potassium and sodium which is the least firmly held.

If the original soil is shaken up with water the liquid will be neutral, because the clay-humus colloid, although it contains acid-forming hydrogen ions, is quite insoluble. If, now, the soil is shaken with a solution of a *neutral* salt such as ammonium sulphate, the liquid will become *acid* because of the formation by cationic exchange of soluble sulphuric acid. This kind of acidity, induced by treating soils with neutral salt solutions, is known as 'exchange acidity.

The exchangeable cations can be varied at will by treatment with an appropriate salt or acid, and it is quite easy to make calcium, sodium, hydrogen, etc. soils in which all the exchangeable cations adsorbed on the colloids are of one kind. Such 'homionic' soils show very marked differences in physical properties according to the nature of the exchangeable cation, and suggest—what is, in fact, the case—that by altering the exchangeable cations the chemist or farmer can exert a far-reaching control over the properties of a soil.

In calcium soils the colloids are thoroughly flocculated and the soils are easy to work, and tend to be permeable and, if other conditions are favourable, to form a crumb structure. Many of the properties most desired in agricultural soils are associated with exchangeable calcium. Sodium soils, on the other hand, have all the most undesirable properties. The colloids are highly dispersed and produce slimy unworkable muds when wet, and equally unworkable hard clods when dry. Sodium soils do not occur in Britain, but are quite common in some arid and semi-arid regions. They are useless as such, but can be reclaimed, and may then become very fertile, by treatment with gypsum (calcium sulphate), whereby the sodium is exchanged for calcium. Magnesium soils are rather like sodium soils, but not quite so bad. Potassium soils are almost or quite unknown in nature, but can be prepared artificially and are similar to sodium soils. Hydrogen soils are physically similar to calcium soils, but owing to the high acidity associated with exchangeable hydrogen they are chemically and biologically poor, and tend to be much less fertile and not to form a crumb structure as readily as the calcium soils.

As would be expected, practically every soil contains a mixture of exchangeable cations. In almost all agricultural soils the predominant exchangeable cation is calcium. In many cultivated British soils calcium accounts for 80–90 per cent, magnesium for 5–10 per cent, potassium for 2–3 per cent and sodium for 0–5 per cent of the total exchangeable cations, excluding hydrogen, absorbed on the soil colloids. Exchangeable hydrogen is negligible in many agricultural soils, but may amount to 80 per cent or more of the cations in acid, moorland or forest soils (such as podzols).

One of the main purposes of applying lime—a term including calcium oxide (quicklime), calcium hydroxide (slaked lime) and calcium carbonate (limestone)—to the soil is to keep up the proportion of calcium in the exchangeable cations. It will not keep up by itself since the calcium, being exchangeable, will exchange with any other cation which comes along, such as hydrogen in acidulated rainwater, ammonium, sodium and potassium in fertilisers. Exchangeable calcium is necessary not only to preserve desirable physical, chemical and biological properties in a soil, but also to serve as a source of calcium as a plant nutrient. The main loss of exchangeable calcium, however, comes from exchange with the cations of fertiliser salts. When ammonium sulphate is added to soil the ammonium is adsorbed by the clay and humus colloids, releasing calcium which combines with the remaining sulphate radicle to form calcium sulphate which gets washed out of the soil and lost. Subsequently the ammonium is oxidised by bacteria to nitric acid, a source of hydrogen ions which take the place of ammonium in the colloid complex. As a result of using sulphate of ammonia, therefore, calcium is lost and the soil becomes acid, two associated conditions which can be remedied by liming.

A similar result follows the use of potassium fertilisers. Potassium enters the colloid complex in exchange mainly for calcium, which is lost. Some of the exchangeable potassium is subsequently utilised by plants as a nutrient, but not all, and only slight acidity develops from using potash fertilisers. When fertilisers containing sodium are used, such as nitrate of soda ($NaNO_3$), the sodium enters the colloid complex and the soil may acquire some of the stickiness of a sodium clay that makes it difficult to work. The exchangeable sodium will also tend to produce an alkaline reaction.

It will be seen, therefore, that the continued use of one fertiliser or of ' unbalanced ' fertilisers may have harmful consequences to the soil by upsetting the balance of exchangeable cations. The two principal nitrogenous fertilisers, sulphate of ammonia and nitrate of soda, both cause loss of exchangeable calcium, but have opposite effects on the soil reaction. With nitrate of soda there is a direct exchange of calcium with sodium which produces alkalinity, whereas with sulphate of ammonia there is first an exchange of calcium with ammonium followed, when the ammonium is nitrified, by an exchange with hydrogen, which is a producer of acidity.

The exchangeable cations which are also plant nutrients are calcium, potassium, magnesium and ammonium, though most of the ammonium is converted to nitrate before it is absorbed by plants. When the cations are present in the soil or applied in fertilisers as soluble salts they are readily available to plants, but in ordinary circumstances only very small quantities of soluble salts occur in soils, and most fertiliser salts quickly undergo transformation to insoluble compounds. The exchangeable cations on clay and humus are probably the main source from which plants get their requirements of calcium, potassium and magnesium. The colloids act as a bank or store for these nutrients. The absorbed cations are protected from being washed out of the soil, but they gradually dissolve in the soil water (soil solution) where plant roots can get them. According to the laws of physical chemistry an equilibrium is maintained between the amount of cations taken out of the soil solution and the amount released from the colloid complex. A large body of evidence has accumulated recently that suggests that plant-root hairs can take up nutrients directly from the colloid complex without the intervention of the soil solution ; the root hair also contains an exchange complex, and a direct exchange takes place when the soil colloids and roots come into contact :

Whether or not and in which direction an exchange takes place depend on the relative concentrations of the cation on the two exchange complexes. If there is a greater concentration of, say, potassium in the root hair than in the soil complex, the root may actually lose this nutrient to the soil. It should be remembered that the soil is not there merely to feed the plant ; the plant is also there to feed the soil, and if the soil is the hungrier the plant may have to disgorge itself to feed it.

One of the most useful tasks of soil chemistry is to maintain a desirable balance of exchangeable cations in the colloid complex. In the current jargon this means ' maintaining the calcium status,' since the other exchangeable cations will usually look after themselves. The faults of a failing calcium status can be remedied by liming, and an empirical quantity known as the ' lime requirement ' of a soil has been invented and can be determined in a laboratory, and enables the chemist to state how much lime must be applied to bring the calcium status to the optimum.

SOIL REACTION

The main symptom of poor calcium status is soil acidity, which would not have required much technical explanation, since everybody knows what acidity implies, were it not for the fact that the invariable custom is to express soil acidity in terms of pH.

pH is defined as the negative index of the logarithm of the hydrogen-ion concentration. In pure water (H_2O) about one part in ten million is dissociated into hydrogen ions (H^+), carrying a positive electrical charge, and the same proportion into hydroxyl ions (OH^-) carrying a negative charge. The concentration of H^+ (or OH^-) in water is therefore 10^{-7}, the index of its logarithm is -7, and its negative index 7. The pH of pure water is 7.

Now the characteristic of acids is that they dissociate into hydrogen ions when dissolved in water. Strong acids such as hydrochloric acid (HCl) dissociate completely, weak acids such as acetic acid dissociate incompletely. If one part of HCl is dissolved in 1,000 of water, dissociation of the HCl will be almost complete and the concentration of H^+ will be 1 in 1,000 or 10^{-3}, and its negative logarithm and pH will be 3. The greater the acidity the lower is the pH. If instead of an acid a strong alkali such as caustic soda (NaOH) is dissolved in water, the soda will dissociate into OH^- (hydroxyl) ions which give alkalinity to the liquid. We do not express alkalinity in terms of pOH, but we can and do express it in terms of pH by making use of the fact that in all aqueous solutions, whether acid, alkaline or neutral, the product of the concentrations of H^+ and OH^- is constant: 10^{-14}. Thus in pure water, as we have already stated, the concentration of both H^+ and OH^- is 10^{-7}, and the same is true of any neutral solution. In an acid solution of pH 3 the concentration of H^+ is 10^{-3} and of OH^-, therefore, is 10^{-11}. In a correspondingly strong alkaline solution, in which the concentration of OH^- is 1 in 1,000 or 10^{-3}, the concentration of H^+ will be 1 in 100,000,000,000 or 10^{-11}, and the pH will be 11.

The lay reader should not trouble himself unduly about the chemical significance of pH beyond understanding that pH 7 means neutrality, pH values below 7 signify increasing acidity, and pH values above 7 increasing alkalinity. It should be noted that pH5 signifies an acidity, as measured by H^+ concentration, *ten times* as great as pH 6, but the effects which matter on plants of a soil pH 5 are nothing like ten times as great as those of a soil pH 6. In Britain soil acidity is seldom a serious factor at pH values above 5·5, and most agricultural crops thrive best in slightly acid soils (pH 6–6·5). In soils more acid than pH 5 various troubles, especially low yields, can be expected, and crop cultivation becomes very restricted in soils below pH 4·5, though forestry with conifers can be successfully carried on on more acid soils.

Most crops can be grown over a fairly wide range of soil reaction or pH, and the range varies considerably between species. Soil reaction is only one of a very large number of interacting factors affecting plant growth, and it is as meaningless to say that a crop does best at, say, pH 6·5 as it would be to say that it does best in soils containing 20 per cent of moisture. Nevertheless there is a lower limit of pH (i.e. a higher limit of acidity) below which a crop grown under the conditions prevailing in Britain is likely to fail more or less completely. Davies (1938) gives the following limits of acidity for the growth of different crops.

TABLE II—SOIL-ACIDITY LIMITS OF COMMON CROPS

	pH		pH
Alsike	5·6	Swedes, cabbages, savoys	4·9
Red clover	5·5	Kale	4·5
Sugar-beet, mangolds, barley	5·3	Ryegrass	4·3
		Oats	4·2
Wheat	5·1	Potatoes	4·0

Among the most acid soils in Britain are those of the Yorkshire moors and of Wareham Heath in Dorsetshire, where pH values in the neighbourhood of 3 and less have been recorded. The Forestry Commission has succeeded in establishing pine and spruce forests in both these areas.

The only alkaline soils in Britain are those containing free calcium carbonate (chalk) derived either from the chalk or limestone rock on which the soil has formed, or from applications of lime. They reach pH values not greater than 8–8·5. Higher pH values, up to 10 or 11, are found in dry-region soils containing sodium carbonate. These highly alkaline soils are useless for agriculture. Not only is the high concentration of OH^- deleterious to plant growth, but it is associated with the presence of exchangeable sodium in the colloid complex, causing extreme dispersion of the soil colloids and producing a very undesirable kind of soil structure. One of the most spectacular applications of the study of cation exchange has been in the reclamation of alkali soils. Treatment with calcium sulphate can effect a complete transformation of their properties. The treatment results in an exchange of sodium for calcium in the colloid complex, and the colloids are transformed from a highly dispersed slimy state to a flocculated crumbly state, and the soil becomes permeable, porous and workable.

In the mildly alkaline calcareous soils of Britain the chief danger is that the alkalinity will render insoluble, and therefore unavailable to plants, compounds of the so-called trace elements—manganese, copper and zinc—which are present in the soil and required by plants in minute amounts, but are none the less absolutely essential to plant growth. Symptoms of deficiency of these elements sometimes appear in plants growing in soils which have been overlimed and made alkaline.

Soil Nitrogen

Practically all the nitrogen in soil is combined in humus and plant residues in which form it is unavailable. It does not become available until it has been converted by bacterial action into nitrate (NO_3) in which form it is, except when the soil is very dry, immediately taken up by any plant which happens to be growing in the soil. Plants which are not fed with nitrogenous fertilisers obtain a steady supply of available nitrogen by the decomposition and nitrification of soil organic matter that in a a fertile soil proceed actively throughout the growing season. In certain conditions, for example, waterlogged conditions, decomposition is impeded and nitrification does not occur. Plants grow poorly and exhibit a yellowish appearance which is a symptom of nitrogen shortage, caused not by the lack but by the unavailability of nitrogen in the soil.

The actual progress of nitrogen from the atmosphere, where all soil nitrogen orginates, into the soil, thence into the plant and back into the soil or atmosphere is known as the nitrogen cycle, and is described more fully in Chapter VI. Here it is only necessary to remark that for all practical purposes nitrogen which is not in the form of nitrate and possibly ammonia (NH_3) is unavailable to plants. In their transformation into nitrate, organic nitrogenous compounds are first broken down by organisms to ammonia compounds, these are oxidised by micro-organisms to nitrites (NO_2), which are immediately oxidised by other micro-organisms to nitrates. The quickest of these stages is the last—nitrification—and very little ammonia and no nitrite (which is poisonous to plants) are normally found in soils, though there is ample evidence that they are formed. Ammonia is so transitory in soil that it is still debatable to what extent it can be used by plants as a nutrient. But it has been suggested that the reason why partial sterilisation of soil (p. 95) produces such dramatic effects on plant growth may be that the process, by killing off the nitrifiers but not the ammonifiers, allows ammonia to accumulate and enables certain kinds of plants to absorb their nitrogen in an exceptionally palatable form.

The making available of soil nitrogen is an entirely biological process carried out by animals, worms, insects and micro-organisms, that the chemist can only control indirectly and to a limited extent by adding to the soil substances, of which lime is the most common, which will stimulate or repress some or other of these organisms. Because the reactions of nitrogen in soil are all biological, chemists have not attempted to unravel them in detail, but are content to record their results which are fairly simple, namely, that atmospheric nitrogen is fixed by bacteria which build it up into body protein, which is ultimately broken down to ammonia and oxidised to nitrate by other bacteria.

SOIL PHOSPHORUS

The phosphorus compounds of the soil and the phosphorus nutrition of the plants have presented soil chemistry with many baffling problems. The chemistry of soil phosphorus, one of the most important plant-nutrient elements, cannot be made to appear simple, and the more we learn of it the more complicated it all becomes. Plants take up phosphorus from the soil in the form of phosphate (PO_4), and the cardinal fact is that in whatever form phosphates are applied to the soil they rapidly become insoluble and mostly unavailable to plants. In virgin soils all the phosphate was originally present as insoluble phosphatic minerals such as apatite, and only became very slowly available to plants by prolonged weathering. When once phosphate had been absorbed by plants and had become included in the biological cycle, however, it was very carefully looked after so that in time a sufficient amount accumulated to satisfy the requirements of a complete plant cover.

Phosphate is ' looked after ' in nature by never appearing in the soil in a soluble form except in very small quantities. It is thus protected from being washed out of the soil, and it is indeed very rarely that any phosphate is found in drainage waters even after the most torrential rain. Even when a very soluble phosphatic fertiliser like superphosphate is applied to a soil, nearly all the phosphate is quickly rendered insoluble (it is said to be ' fixed ') and unavailable to plants. It is for this reason that the normal recovery, in a crop, of phosphate given as fertiliser is of the order of only 10–20 per cent. The unlikely discovery of some means of preventing phosphates from being fixed in soil would greatly increase the productive capacity of almost every soil in the world. It may be noted that what the agriculturist would like from the chemist with regard to *nitrogenous* fertilisers is one that is less soluble and more slowly acting than those now made—the opposite of what is wanted of phosphatic fertilisers.

Phosphorus is present in soils as (1) completely insoluble inorganic materials of the apatite type (complex calcium phosphates) ; (2) very insoluble iron and aluminium phosphates ; (3) slightly soluble dicalcium phosphate ($CaHPO_4$), which is probably available to plants, but is unstable and changes over to less soluble forms ; (4) ' exchangeable ' phosphate, a negatively charged anion (PO_4) which seems to be adsorbed on the soil colloids in an analogous way to the positively charged cations and can be exchanged by other anions such as sulphate (SO_4) ; (5) organic compounds present in plant residues and humus that may be decomposed by biological action releasing phosphate in a plant-available form.

Of these forms the ones useful to plants are the unstable dicalcium phosphate, possibly the exchangeable phosphate and the phosphates released by decomposition of organic matter. It is probable that wild vegetation gets most of its phosphate from the last. It is found that soil phosphate is most available at a slightly acid reaction of about pH 6 ; with greater acidity there is a tendency to form insoluble iron and aluminium phosphate, and with greater alkalinity insoluble apatite-like compounds. Stimulation of the decomposition of organic matter is one way of increasing the supply of plant-available phosphate. Good results have been obtained by ' composting ' superphosphate with farmyard and other organic manures, whereby the phosphate is first taken up in unavailable organic combination, and subsequently released in available form during decomposition of the compost. When superphosphate is applied directly to soil, some, of course, remains available long enough to be absorbed by growing plants, but recent work (Spinks and Barber, 1948), using radioactive phosphate whose fate in the soil or into the plant can be followed, has shown that fertiliser phosphate is utilised mainly in a few weeks after its application, and that thereafter plants get their phosphorus not from the fertiliser but from the soil, probably by the decomposition of organic matter.

Plants vary considerably in their ability to absorb phosphate from the soil, and some can take it up in forms which are unavailable to others. Lupins, especially, and several other legumes have been found to be very good extractors of phosphate from insoluble compounds ; cereals are bad extractors. Green manuring with good phosphate extractors, especially if they have deep roots which can bring up phosphate from the subsoil, is a useful way of adding to the amount of phosphorus in the biological cycle, because, when the green manure is ploughed in and decomposes, the phosphate extracted by the roots becomes available not only to plants which are good, but also to those which are bad, phosphate extractors.

In comparison with the soils of some other lands, those of Britain are neither rich nor poor in phosphorus, but like those of all agriculturally highly developed lands they do not contain nearly enough in available form for the heavy crops now obtainable. They vary much in their phosphate-fixing capacities. Soils with low phosphate-fixing capacity will, other things being equal, respond to phosphatic fertilisers more readily than those with high-fixing capacity. High fixing capacity is associated with extreme acidity or alkalinity, and the presence of oxides of iron and aluminium which occur in many soils, especially in the tropics, where little response is obtained to phosphatic fertilisers even though the soils are generally deficient in phosphate.

Soil Potassium

Plants appear to derive most of their potassium from the exchangeable potassium adsorbed on the clay-humus complex, but not all the exchangeable potassium is available. The amount of exchangeable potassium does, however, give a fairly reliable indication of a soil's probable response to potassic fertilisers. Some potassium-containing soil minerals, such as muscovite, weather readily, and may provide steady sources of plant-available potassium. Many plant residues also contain potassium in a soluble and therefore immediately available form. Applying a surface mulch of straw or similar material is a very effective way of increasing the available-potassium content of the soil ; the extra potassium can only have come by being washed out of the mulch. Humus is not a source of potassium.

Trace Elements

The elements nitrogen, phosphorus and potassium are the 'big three' in plant nutrition. They are required by all crops in much larger quantities than other nutrient elements obtained from the soil, and deficiencies of them, resulting in low yields and maldevelopment, are very common. They can usually be overcome by applying fertilisers or manure, provided the soil is in a physical condition in which plants can use the nutrients (e.g. is adequately drained). There are other elements necessary for plant growth that are required in smaller quantities, among them calcium, magnesium (a constituent of chlorophyll), sulphur and iron, deficiencies of which occasionally occur, but most soils contain ample quantities in available form to satisfy a crop's modest needs.

A third group of elements, including manganese, copper, zinc and boron, and known as 'trace elements', are required in such minute traces that their presence and importance in plants have only recently been discovered by refined methods of analysis. They are, nevertheless, quite essential for normal plant development, although if they are absorbed by plants in more than necessary quantities they may be toxic. These elements are present in most soils, but only in small quantities, and conditions may arise when there is not sufficient in ' available ' form to satisfy even the minute requirements of a crop. The commonest condition conducive to trace-element unavailability is an alkaline soil, since most of the compounds of the elements that occur in soils are insoluble in alkaline solutions.

Overliming until the soil is alkaline is a frequent cause of trace-element deficiency which shows itself in various forms. Grey speck of oats and marsh spot of peas are manganese-deficiency diseases, and heart rot of sugar-beet is a boron-deficiency disease, that can usually be cured by making the soil slightly acid or, more easily, by applying very small quantities of a salt of the appropriate element to the soil, or by spraying a salt on to the foliage of the crop.

These diseases are of little economic importance in Britain, but there are countries where they or other trace-element-deficiency diseases are serious. There are peat soils in Holland, Germany and the United States where crops cannot be grown without small applications of copper salts, and serious deficiencies of all the trace elements have been reported from parts of Australia or New Zealand. In Australia large areas of formerly useless land have been made productive by using fertilisers containing small quantities of copper and zinc.

Cobalt is an element which is not necessary to plants, but is necessary to animals in minute quantities, and if a soil lacks it pasture growing on the soil will lack it also (without showing any deficiency symptoms), and stock feeding on the pasture will suffer. The wasting disease of cattle and sheep, known as 'pine' in Britain, 'coast disease' in Australia, 'bush sickness' in New Zealand and 'salt-sick' in America, has been traced to a deficiency of cobalt in the soil and hence in pastures. It has been found in Britain in Devon and Cornwall on granite soils, and in the Cheviot Hills on several geological formations. It can be completely cured by adding a few pounds of a cobalt salt to an acre of land.

CHAPTER IV

SOIL WATER

EVERYTHING that is not solid in a soil must obviously be either liquid or gaseous, which means, in effect, that all the spaces and pores between and connecting the solid particles are filled with either soil moisture or soil air, both being slightly different from moisture and air as generally understood. The total volume of moisture plus air in a soil is thus limited to the ' pore space,' and lack of sufficient moisture or air is the chief soil factor limiting plant growth. The more moisture there is the less room there is for air, and the more air there is the less room there is for moisture. Since both are equally required for the healthy growth of plant roots the optimum state for a soil is one in which both air and water are present, even when growing crops are not getting all the water they could use advantageously. Once again, we find that this optimum state is most easily attained when the soil has a crumb structure, the granules absorbing water and at the same time allowing free drainage so that air can enter the soil pores.

Moisture is held by a soil with varying degrees of tenacity, from the loosely held moisture which drains out of a saturated soil to the firmly held moisture that can only be expelled by heating. Soil exerts a kind of suction on water that is greater the less water is present. Plant roots also absorb water by a similar kind of suction, and can only absorb it when the suction by the root is greater than that by the soil. Hence at low soil-moisture contents when suction by the soil is high the water in the soil is unavailable to plant roots.

Moisture Constants

It is convenient to divide soil moisture into arbitrarily defined classes distinguished by the relative tenacities with which the moisture is held by the soil. The most firmly held water of all is the *chemically combined water* which is part and parcel of the soil minerals, and can only be expelled by heating to red heat. Actually, since this water is chemically combined it does not moisten the soil at all, and can be ignored as ' moisture.' A soil which has been heated for some time at a few degrees above boiling-point will contain no free water and is known as *oven-dry*. When an oven-dry soil is left in contact with moist air it will absorb *hygroscopic moisture*, the amount depending on the moistness of the air, the temperature, the physical condition of the soil, etc. The soil is then *air-dry*. The percentage of moisture absorbed by a soil from air saturated with water vapour is called the *hygroscopic coefficient*. Hygroscopic moisture is unavailable to plants.

When a sample of soil is thoroughly wetted and put in a position in which it can drain freely, some water will drain away by gravitation (*gravitational water*), but even when drainage has ceased completely a good deal of water will remain in the soil, particularly if it is clayey. The water that remains is the *capillary moisture* (or, more accurately, the capillary plus hygroscopic moisture), and it is held in the finer pores of the soil by capillary forces which are set up by surface tension between the water and the walls of the fine pores, tubes and capillaries of the soil. The smaller the pores the stronger is the capillary force, and when the force is stronger than gravity it will suffice to prevent drainage in a freely draining soil. Capillary moisture acts as a kind of reserve of water, protected against drainage, available to crops if not too strongly held, and capable of limited movement within the soil. In a sandy soil the pores are fairly large, and the soil will drain fairly completely ; in a compacted, structureless clayey soil all the pores will be very small, and a considerable proportion of all the water the soil can hold will be held by capillary forces. In a soil with crumb structure, and therefore containing both small and large pores, the water in the minute pores within the granules is held strongly by capillarity, while the water in the larger spaces *between* the granules is partly capillary and partly gravitational. Thus a sandy soil is liable to be overdrained, structureless clayey soil to be underdrained and a structured soil to be both well drained and water-retentive.

The amount of water retained by a previously saturated soil when free drainage has ceased is known as the *moisture-holding capacity* or the *field capacity*. It can vary from very little up to 60 per cent or more of the weight of the soil, according to the texture and structure of the soil. If no evaporation or transpiration by plants took place a soil would remain at field capacity indefinitely. When the soil is completely saturated with water and contains no air it is at its *maximum moisture capacity*. It is then waterlogged.

The *wilting point* is the percentage of water in a soil at which plants wilt for lack of moisture and, having wilted, will not recover when more water is added to the soil. For any one soil it is approximately the same for many plants, and it represents the point below which soil moisture is unavailable to plant roots. As with other moisture constants it varies with soil texture and structure. Any water present above the wilting point may be regarded as ' available ' to plants. The wilting point might occur at, say, 5 per cent moisture in a light sand, and at 15 per cent moisture in a heavy clay. It could be even higher in some kinds of peat which had a high attraction for water due to the presence of much humus. Indeed, one reason why heather dominates some of our often soaking wet peat areas is because the plant can live on much less water than most others.

The *moisture equivalent* is an arbitrary but useful constant for characterising soil-moisture behaviour. It is the percentage of water remaining in a soil which has been subjected to a centrifugal force 1,000 times the force of gravity. This force removes all the capillary water except that held in the finest pores and capillaries and the so-called *imbibitional water* which is that absorbed by the soil colloids when they swell on moistening. This equivalent is rather less than the field capacity.

The three constants hygroscopic coefficient, wilting point and moisture equivalent do not necessarily have the same relative values for all soils, but in the order given they represent increasing degrees of moistness. This means that the suction power by which plant roots withdraw water from the soil is greater than 1,000 times the force of gravity (at which moisture is held in the soil at the moisture equivalent) and less than that with which an air-dry soil holds small quantities of moisture.

CAPILLARY POTENTIAL

In a soil at or below field capacity, movement of water in any direction is slow and restricted. It takes place mainly by capillarity, as a thin film moving over the surface of the soil particles or through the fine pores and capillaries. In certain circumstances it can also take place in the form of vapour, by distillation. Capillary movement of water in soil is governed by the same forces as govern the movement of water through a porous brick when one end is immersed in water. Water will move through the pores of the brick from the wet end towards the drier end, and the degree of wetness of the brick will decrease with the distance from the wet end. The general rule is that in a porous system moisture will always move by capillary suction from a wetter to a drier part, until the suction force is counterbalanced by other forces, of which the main one in soil is gravity. This movement of moisture has been compared with the movement of electricity through a conductor from a point of high to a point of lower electric potential, and the comparison has led to the introduction of the concept of *capillary potential* to define the suction force by which moisture is held by soil particles. Dry soil has a high, and wet soil a low, capillary potential, consequently moisture moves from a low to a high potential, and will continue to do so until the whole system is at the same moisture potential. This is not necessarily the same as having a uniform percentage of moisture throughout the system. At the same moisture percentage a sand will have a lower capillary potential than a clay soil, and if the two are brought into contact moisture will move out of the sand into the clay.

The capillary potential rises rapidly as the moisture content decreases below the wilting point, at which it is about 16 atmospheres. At the hygroscopic coefficient it rises to about 1,000 atmospheres. The range of moisture content, on the other hand, may be 5 per cent at the hygroscopic coefficient, 10 per cent at the wilting point and 25 per cent at the moisture equivalent, at which the capillary potential is about 1 atmosphere. It is difficult to express on a curve the relationships between the two variables capillary potential and moisture content when the numerical ranges are so different, and to overcome this difficulty Schofield (1935) has introduced the term pF, analogous to the pH used to define soil reaction (p. 43), to define the suction force or capillary potential of a soil. The force with which water is held by soil is equal to the force needed to extract the water from soil. Thus if moist soil is submitted to a suction or negative pressure of 1 atmosphere (on a vacuum filter, for example), the water remaining in the soil will be held by a suction force also equal to 1 atmosphere's pressure. This pressure is about 76 centimetres of mercury or 1,000 centimetres of water. Schofield defines the pF of a soil as the logarithm of the suction force, expressed in centimetres of water, with which the soil moisture is in equilibrium. A capillary potential of 1,000 centimetres of water would thus be expressed as pF 3 (log 1,000), which is, in fact, the pF of a soil at the moisture equivalent. The pF at the wilting point is 4·2, in the air-dry state 6, and in the oven-dry state 6·9—equivalent to a pressure of nearly 10,000

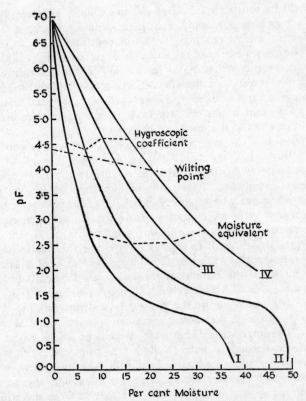

Fig. 1. Relation between pF, moisture content and moisture constants (after M. B. Russell, 1940)

atmospheres or of a column of water 60 miles high. The pF at field capacity is less definite, but is usually about 2·5 of less.

    Fig. 1 shows pF-moisture curves of four American soils. At equal moisture contents the pF (a measure of the ' wetness ' of a soil) increases as the texture of the soil gets heavier. I is a sand, II a sandy loam, III a silt loam and IV a silty clay.

The various moisture constants, determined independently, are :

	I	II	III	IV
	%	%	%	%
Air-dry moisture	0·81	1·67	2·47	4·80
Hygroscopic coefficient	3·41	6·93	10·4	16·1
Wilting point	3·7	7·2	12·7	20·6
Moisture equivalent	7·7	15·9	24·4	31·2

It will be seen from these figures and from the curves that the pF values for each constant are approximately the same, though the corresponding moisture contents vary greatly.

MOVEMENT OF SOIL MOISTURE

(i) *As Liquid*

When a large quantity of water is added to a soil—by irrigation, for example—the water does *not* distribute itself evenly through the soil mass. The top layer will be saturated to field capacity, and only then will the underlying layer be wetted. It is for this reason that a heavy rain following a long dry spell often seems to have penetrated such a disappointingly small distance into the soil, and that there is a perfectly sharp boundary between the wetted soil and the underlying dry soil. It is, of course, a very good thing that water does behave like this, because it means that most of the rainwater remains within the root zone ; if added water distributed itself evenly throughout a soil, a deep soil would seldom get adequately wetted to grow a crop.

When soil is saturated above field capacity the excess water can move freely by gravitation, but below field capacity there is hardly any movement of liquid water at all, except into

growing plants. It has already been stated that there is a
tendency for moisture to move from a wet (low pF) region
to a dry (high pF) region, but in fact this movement is so slow
that it can be ignored, except when it is movement from a
permanent water table or by seepage from a pond or river—
that is, from a *continuously* saturated to a drier region. Then
a moisture gradient is set up by capillary movement of water,
the soil getting drier as its distance from the saturated zone
increases.

It was formerly thought that plant roots could get moisture
by capillary movement upwards from the underground water
table, but it has recently been found that this does not happen
unless the water table is quite near the surface. It has been
shown conclusively at Rothamsted that water will not move
upwards by capillarity in a clay-loam soil more than about
$2\frac{1}{2}$ feet from a permanent water table ; for a coarse sand the
maximum rise is about 1 foot. Again, it was once thought
that one of the objects of surface cultivation was to conserve
soil moisture by producing a ' dust mulch ' which would
destroy the soil capillaries and thus prevent subsoil moisture
from reaching the surface and being exposed to evaporation.
This effect is, however, illusory. Liquid water—except
gravitational water—scarcely moves at all in soil. When a
plant root absorbs water from a soil in its immediate vicinity,
more water will not come to it by capillary attraction from
a wetter part of the soil ; the root will have to grow and go
after the water it needs.

(ii) *As Vapour*

If the vapour pressures at two points in a soil are different,
water vapour will be transferred from the point with the
higher to the point with the lower pressure. At the same
temperature, however, vapour pressure is practically constant
and independent of the wetness of the soil at all moisture
contents above the wilting point. But if the temperature varies,
the vapour pressure will rise with the temperature, and water

will tend to distil as vapour from a point at a higher temperature and to condense as liquid at a point at a lower temperature. When the surface soil is warmer than the subsoil, water will distil downwards, and upwards when the subsoil is warmer, but only if there are air-filled spaces through which the vapour can pass. It is not known how much moisture moves in this way, but the net movement either upwards or downwards is probably not very great. Downward distillation may, however, account for the drying-out of surface soils to a depth of several inches during the summer, as evaporation alone should dry only the thin layer through which water will move by capillarity.

The amount of water draining from a soil sometimes increases with an increase in the air temperature, even when there is no rain (Roseau, 1948). The extra heat causes the water in the surface soil to vaporise and distil downwards to cooler layers below. Cultivation encourages loss of water by evaporation, but in uncultivated undisturbed soils it is possible that in hot dry periods the main movement of water through soil may be not upwards, but downwards, by distillation.

DRAINAGE

After a fall of rain soil will become wetted to field capacity from the surface downwards. As already mentioned, the water will not spread itself evenly throughout the soil, but will moisten one layer up to field capacity before moving on to the next. When the rainfall (or the irrigation) is sufficient the whole soil will be saturated to field capacity from the surface down to the permanent water table, which is the level at which ground water appears, and the depth at which water is found in wells. It is also the depth at which the soil is completely and permanently saturated with water, and it has a very great influence on the productivity of soil. Where it is near the soil surface, as in a low-lying meadow adjoining a river the level of which is the water table, plants are liable to suffer from excess of moisture because relatively little water is needed to saturate the shallow depth of overlying soil. Where it is very deep and right out of the range of plant roots the ground water cannot serve as a reservoir from which plants can get water, although some water can move upwards by capillarity for 2–3 feet from the water table. This damp zone, moistened from the water table by capillarity, is called the *capillary fringe*. Fig. 2 shows the root systems of barley plants grown with water tables maintained at the depths indicated by the horizontal lines beneath the roots, and it will be seen that in every case except the last one on the right (where some disturbance occurred) root growth reached down to just above the water table.

It is obvious that soil below the water table is useless as a source of both plant nutrients and water. Under most agricultural conditions there are appreciable fluctuations in the water table in the course of a year, ground water tending to accumulate in wet seasons and to be dissipated in dry seasons. A mottled grey-and-brown layer (known as the ' gley horizon ') is sometimes found in a soil, and indicates the zone within which the water table fluctuates. When, in wet periods, the water table rises to the top of the gley horizon, the soil below the water table is waterlogged and, in the absence of air, brown ferric-iron compounds in the soil are converted into grey ferrous-iron compounds. When the water table falls, air gets in and the ferrous compounds are re-oxidised to brown ferric compounds. But if there are any patches of impermeable soil where drainage is impeded and the soil remains saturated even after the water table has fallen, the grey colour persists, giving rise to the mottling of the gley horizon. It will often be found that the brown patches are composed of noticeably sandier soil than the grey patches, or occur round old root channels from which water will drain easily. Mottling in the subsoil is an almost infallible sign of impeded drainage (Plate 3a).

If the water table rises to within the root zone, especially during the active growing season, the roots may be killed by

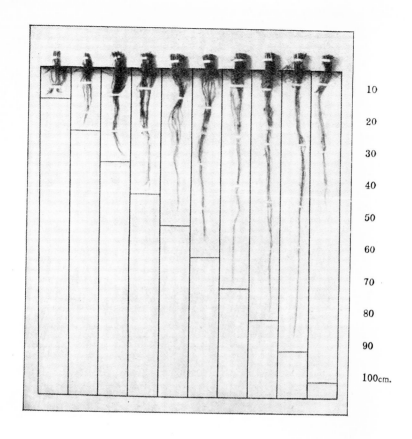

10

20

30

40

50

60

70

80

90

100cm.

Figure 2   The effect of depth of water table on the development
of crop roots

lack of air and a fall in yield will result. In Table III are shown the relative yields of barley obtained when the water table, normally kept at 80 cm. depth, was raised for a fortnight to depths of 10, 25 and 50 cm.

TABLE III—PERCENTAGE YIELDS OF BARLEY OBTAINED WITH WATER TABLE AT 80 CM., AND RAISED FOR A FORTNIGHT TO 10, 25 AND 50 CM.

Periods during which the water table was altered		Water table at 80 cm., but raised once to			Water table at 80 cm. all the time
		10 cm.	25 cm.	50 cm.	
June 10 to 25	Grain	21	30	57	100
	Straw	23	38	57	100
June 20 to July 5	Grain	26	46	76	100
	Straw	26	47	70	100
July 1 to 15	Grain	20	29	39	100
	Straw	33	50	51	100
July 10 to 25	Grain	33	36	45	100
	Straw	37	45	41	100

A water table may be produced by the presence in the soil of an impermeable layer through which water percolating from the surface cannot pass. Such a water table is ' temporary,' but the water may disappear through evaporation and transpiration, and it cannot be replenished from the ' permanent' lower-lying ground water, connection with which is cut off by the impermeable layer. In the humid climate of Britain the majority of agricultural soils have the water table quite close to the surface. The purpose of drainage is to lower the water table, whether temporary or permanent, and thus increase the depth of soil effective for plant growth. Paradoxically, increasing the depth of soil by draining away excess of water may mean that more water is put at the disposal of the crop, and that by allowing roots to penetrate to a greater depth crops are better able to stand a drought than in undrained soil.

Drainage is done by providing channels down which water can flow easily, and its effect is to lower the water table to the depth of the drain. The channel can take the form of an open ditch or a covered tile or mole drain. Such drains will remove all excess water over field capacity from saturated soils, and in so doing will cause a certain amount of shrinkage in the soil in the immediate neighbourhood of the drain, thus producing subsidiary channels through which water farther away from the drain can be removed. The immediate effect of drainage is to let air into otherwise waterlogged soil, but it has other associated physical advantages, among which is an improvement in the structure of heavy clay soils, as a result of the partial drying out and flocculation of the clay. At the same time a certain loss of plant nutrients (chiefly nitrate) from the soil takes place as a result of drainage, though this is not generally serious except when fertilisers have been applied. But if the soil needs draining and is not drained the fertiliser would not have been of much use anyway.

In arid regions where irrigation is necessary, drainage is just as important as irrigation. The first thing to do in planning an irrigation project is to see that the drainage will be adequate, otherwise irrigation may be not only useless, but positively and irreparably harmful. If the excess water cannot drain away, it will form an underground lake, the permanent water table will rise and finally the lake will appear on the surface and the soil will be completely and permanently waterlogged. There is danger of this happening in some of the greatest irrigation works in the world.

Attempts have been made to determine what is the optimal soil-moisture content or rather, the optimal pF, for crop growth, but there probably is no clearly defined optimum. When plants are grown in different samples of the same soil maintained at different moisture contents, plant growth at first increases with increasing moisture content, reaches a maximum, then falls sharply to almost zero when the soil is completely saturated with water. Under natural conditions, of course, the soil-moisture content is not kept constant throughout the period of plant growth, and it is quite possible that a fluctuating moisture content, by generating an interchange of water and air in the soil pores, may itself benefit plant growth.

The most generally favourable conditions for assuring growing plants an adequate supply of moisture and air are those of a soil with a crumb structure. In times of excessive moisture the excess can drain away through the large interstices between the granules, leaving sufficient of both air and moisture for plants. At field capacity, when surplus water has drained away, there is plenty of water readily available to plant roots, on the surface of the granules. The moisture in the interiors of the granules is very strongly held, and mostly unavailable to plants, but it stays there as a reserve against drought periods, when evaporation will draw it out from the interior on to the surface of the granules where the thirsty plant roots can get it. A crumb structure thus makes the best of extremes of both wetness and dryness as well as of intermediate conditions. The soil type with the best developed crumb structure is the black, humus-rich chernozem of steppe country. Here the greater part of the annual precipitation occurs as snow, in winter. When the spring thaw comes, therefore, almost the whole year's supply of moisture has to get into the soil in the course of a week or so, or else be lost to the following crop. This can be done if the soil possesses a good structure ; it can absorb all the snow water and conserve it against the dry summer months when it will give up the water continuously to the growing vegetation. If, on the other hand, the natural crumb structure of a chernozem is destroyed—by cultivation, say—the soil will be unable to absorb the great quantity of water suddenly produced by the thaw, the water will run off the soil surface, carrying soil with it, and there will not be enough water, and perhaps not enough soil either, left to produce a decent crop.

In Britain the farmer has a more copious and more evenly distributed rainfall than the farmer of the steppe, but he does not know when to expect a wet or a dry season. If he can produce a crumb structure in his soils he will make the best of whatever conditions befall.

CHAPTER V

THE SOIL POPULATION: I FAUNA

' DEER, jumping mice and the oven-bird are denizens of the forest floor by virtue of using it as their substratum, but there is also a host of curious animals which use the forest floor, especially the litter of dead leaves, twigs, branches and fruit parts, as their walls, ceiling and sub-basements. Looked at from the eye level of the cockroach, this litter becomes a several-storey edifice of enormous extent. The various floors are separated by twigs, midribs, petioles, fruit husks, samaras, skulls, elytra and faeces. The lower one descends, the more compact is the structure. The leaves become more fragmentary, the faeces of worms which have come up from the soil, of caterpillars which live in the trees and of the inhabitants themselves, as well as grains of sand brought up by the worms and a heterogeneous assortment of beetle skulls and wing covers, become more abundant. This complex is rendered more intricate by the growth of minute fungus moulds which feed upon the dead leaves and organic refuse, weaving it all into a compact mat by their myriad white hyphae. Thus is the woof woven into the warp of the woodland rug.'

This quotation from the works of an American forest biologist (Jacot, 1935) gives a graphic picture of the activity of the teeming animal life of a forest. Animals are indispensable to the functioning of a stable living community, and much of their function consists of disposing, in various ways, of the output of material produced by the equally indispensable activities of the plant members of the community. In so doing, the fauna becomes one of the most important agents in the formation of soil. It can consist of animals of all sizes, from elephants to protozoa. For our purposes it will be convenient to distinguish the macrofauna, whose members are visible to the naked eye, from the microfauna, whose members are invisible, although the distinction is quite arbitrary.

MACROFAUNA

The larger soil-inhabiting animals occurring in Britain do not have much effect on the properties or development of the soil, though it would be wrong to say that they have no effect. Rodents like rats and mice are seldom present in sufficient quantity to affect the soil, however pestilential they may be to field products. Rabbits can have a profound effect on the soil, sometimes by their burrowing activities, but more frequently by transforming the vegetation, and hence the direction of soil formation. Large areas of Australia have been laid waste by rabbits, which consume all or nearly all the vegetation, and thus not only deprive the soil of its living protection against wind erosion, but also cause breakdown of soil structure and loss of fertility. Similar phenomena can be observed in England on the Suffolk Breckland, where the rabbit is one of the most potent influences, through its effect on the vegetation, in soil formation. The rabbit is a vegetarian, and will settle anywhere where he can get the food he likes, where he is not exterminated by his enemies and where he can burrow.

Moles

In Britain and, indeed, in a large part of the whole world the mole has the greatest effect of all the larger soil-inhabiting animals on soil formation. It is an efficient, though not usually welcome, cultivator. Moles may completely undermine a field, causing a subsidence of some parts of the surface, and a raising of other parts by the material thrown up in mole-hills. The fine crumby structure of the soil of many mole-hills is very noticeable, but it is unlikely that the structure is produced by the action of moles. Mole runs help to aerate and drain the soil, and moles bring up material from the subsoil and mix it with the surface layers. Mixing by moles and rodents can be an important factor in increasing the depth of the humus-containing layers. In many steppe soils, in particular, rodent burrows are very frequent to depths of five feet or more, and their effect in thoroughly mixing the whole soil is very obvious.

The mole lives off insects and, especially, earthworms. A mole will dig out many yards of run with many side branches which, in turn, are branched. The position of the runs can be seen by the molehills which are thrown up frequently along their course as the mole digs. Where conditions are favourable, that is, where there are plenty of earthworms the whole soil may be honeycombed with mole runs. I once lived near a grass tennis court on an acid sandy soil. Every spring, when the last season's markings of the court had long since been washed away by the rain, the position of all the lines could be accurately traced by the molehills that appeared. What had happened was that the chalk used for marking the court had made the soil below the lines, but none other, a suitable habitat for earthworms, and moles had confined their runs to these places only. A large mole population indicates a large earthworm population. It has been calculated (Hoffmann, 1931) that a mole eats one and a half times its weight of food, mostly earthworms, daily. Moles are said to bite off the front segments of earthworms and to store their corpses as a winter food supply. In one mole-run 1,280 worm corpses were found, collected in batches of ten or so, immured in the walls of the run.

Earthworms

Earthworms are the only members of the soil fauna whose ecology has been studied at all, albeit only in outline. In soils in which they thrive they usually form the major part, by weight, of the total faunal population. There are many different kinds of worms, and they behave differently in the soil. Some eat their way through the soil, some squeeze their

way through ; some are humus producers, some humus
destroyers, and most are both producers and destroyers.
Earthworms have been described as the farmer's best friends,
and they are certainly remarkably fine cultivators of the soil,
nevertheless the farmer himself can generally do even more
than earthworms in making a soil fertile for agricultural crops.

In uncultivated soils the actions of earthworms are mani-
fold. They carry dead vegetable matter such as leaves into
the soil, where they consume it along with soil, and excrete
it as humus intimately mixed with the mineral soil. Earthworms, more than any other agency, are responsible for
combining humified plant residues with rock particles and
converting them into what is commonly called soil. Darwin
(1881) estimated that earthworms move 7–14 tons per acre
from the subsoil to the surface in a year, and more recently
Evans (1948) has estimated the weight of casts produced per
acre to be 1–25 tons, and the weight of soil passing through
the alimentary tracts of all the worms present, including those
which do not cast on the surface, to be 4–36 tons per acre
per annum. Earthworms like soils containing lime, which
they require for the functioning of some peculiar ' calciferous
glands' which secrete calcite and play an important part in
neutralising the acids produced in the humification of plant
residues, and in combining the humic matter with clay minerals
to form the ' clay-humus complex ' which is the ultimate
brown stuff usually regarded as soil.

A very good illustration of the effect of worms and other
soil-eating animals on soil is got by comparing the properties
of a deciduous-forest soil with those of a coniferous-forest soil,
especially if we can find the two close to each other, as Müller
(1887) found them in Denmark. Müller noticed that the
deciduous-forest (oak) soil was a *mull*, that is, a brown, homogeneous mixture of humus and mineral matter so intimately
mixed as to be inseparable, whereas the coniferous-forest
(pine) soil consisted of a peaty layer of ' raw humus ' or *mor*
overlying and quite distinct from the bleached structureless

mineral soil (the podzol described on p. 10). The oak soil had a rich faunal population, including earthworms, the pine soil contained few animals and no earthworms. It was the faunal population of the oak soil that had disposed of all the dead leaves and litter, had humified them and mixed them with the main mass of the soil. In the pine soil, where the faunal population was insignificant, humification was done by micro-organisms, especially fungi, and did not go much further than converting the plant material into a peaty mass that retained some of its original plant structure.

Worms are equally effective in distributing bulky organic manures through, and incorporating them in, cultivated soils. It will have been noticed that organic manures disappear more quickly from some soils than from others. Rapid disappearance is usually associated with a reasonable lime content in the soil and a large worm population. If, for example, grass clippings are buried in an acid (lime-poor) soil they will be found a year later a little darker in colour and more humified, but otherwise unchanged. In a neutral (lime-rich) soil containing earthworms the clippings will have disappeared completely.

While some kinds of worms eat their way through soil others push their way through, as moles do. These riddle the soil with small burrows going down into the subsoil where the worm population retires to when the weather is very dry or cold. The burrows serve as channels both for drainage water and for plant roots to grow in. They also help to aerate the soil and thus make it a more suitable habitat for animal life in general. In cultivated soils which are being continuously disturbed these actions are not so important as in uncultivated soils.

Bornebusch (1930) estimated that the weight of worms in an oak-forest soil was equal to the weight of livestock on a well-stocked Danish farm, assuming that the livestock were evenly distributed over the whole farm. This probably means very little, but it does show that worms, which work continuously on the soil, might be regarded as at least as important factors in soil management, at any rate in pastures, as livestock. When grassland is ploughed up the worm population falls rapidly. This is not necessarily detrimental to soil fertility, since much of the beneficial work done by worms in pasture and forest soils can be done by the cultivator himself. He loosens and aerates the land—not so efficiently, perhaps, as the earthworm, but well enough for his purpose. He cannot, however, perform the worm's function, so essential in grassland, of humifying plant residues and mixing them evenly with the soil. Earthworms and other small animals are necessary for this, and, in fact, usually appear on the scene wherever there is organic matter to be humified. Where the farmer does not apply organic manures to his soil the worm population is likely to fall with the organic-matter content of the soil, but that does not matter so much because, with little or nothing to be humified, there is less useful work for worms to do.

A formation of peaty raw humus, similar to that formed in coniferous forests, is often found on very acid grassland in which worms cannot live. In the permanent-grass plots at Rothamsted, in which each plot has had the same fertiliser treatment annually since 1856, the plot which has received sulphate of ammonia only (which has made the soil very acid) has developed a dense peaty sod over the mineral soil, produced by an accumulation of half-humified grass residues which have not, as in the other plots, been mixed in with the soil by earthworms. A plot which has received sulphate of ammonia and lime, which neutralises acidity and encourages earthworms, has not developed a peaty sod.

Earthworms have a very remarkable effect on soil structure, produced, probably, by bringing the three chief ingredients of a soil crumb—humus, clay and lime—into very intimate contact within the worms' guts. Worm excrement has a very stable fine crumb structure, and wormcast soil has been found to have a much higher content of plant nutrients than nonwormcast soil. This does not mean, as has sometimes been assumed, that worms actually increase the amounts of nitrogen, phosphorus and potassium in a soil, but rather that they concentrate these nutrients in their casts at the expense of the rest of the soil, and possibly increase their availability. Nitrification is active in wormcasts.

Introduction of worms into a worm-free soil may, if conditions are favourable, produce an increase in productivity. It used to be thought that the increase was produced by the manurial effect of the worms' bodies, but American experiments (Hopp and Slater, 1949) suggest that it is produced by improving soil structure. In these experiments pairs of barrels were filled with very infertile soil variously treated with fertilisers and mulches, the only difference between members of a pair being that one received living and the other dead earthworms. Very striking results were obtained in the barrels treated with live worms, grass and clover growing luxuriantly compared with almost complete failure in the dead-worm barrels treated with lime and fertiliser. Ants, which happened to invade one of the dead-worm barrels, had the same effect on plant growth as worms. The effect was the result of a great improvement in soil structure. Dead worms did have a manurial effect, but it was not nearly so great as the effect of live worms. Hopp and Slater found that the manurial effect of dead worms was apparent during the summer, whereas the physical structure-forming effect of live worms operated mainly during the autumn, winter and early spring. They suggest that for this reason it may be advantageous to create conditions that will enable worms to over-winter in large numbers. Worms cannot tolerate cold (freezing point or lower), and measures that protect the soil against cold, such as applying a mulch or having a grass cover during the winter, will help their survival.

Earthworms like a warm moist soil and lime, and given these they will generally multiply up to the limit of their food supply, which is plant residues which they will pass through their bodies and turn into soil. Contrary to a widespread belief, they do not mind artificial fertilisers in the amounts usually applied, though naturally they prefer organic manures which they can eat. And it must be remembered that from the farmer's point of view they are not so useful when artificial fertilisers only are used, because their chief value to the farmer is in incorporating organic matter into the soil. Attempts have been made in America to commercialise earthworms and even to breed special, super-efficient humus-producing varieties which are claimed, perhaps with justice, to have been introduced into infertile land and to have made it fertile. It is conceivable that in arid country there may be no earthworms in the vicinity from which a population could be built up when the soil is irrigated and the aridity removed. In such circumstances artificial introduction of worms may be necessary, but in a country like Britain it can be taken for granted that if earthworms can live in a soil they will be there already, and if they cannot attempts to introduce them will fail. The presence of earthworms is an indication that such important factors as soil reaction and decomposition of organic residues are favourable for crop production, but large crops can be obtained in their absence, and the farmer should remember that his prime job is to produce crops, not worms.

Ants

Ants are among the commonest and most widespread species of insect that inhabit soil. More than 5,000 kinds of ants are known. They live in highly organised communities in nests that may be underground or raised above the soil surface in the form of anthills made of soil and plant remains collected from the neighbourhood. An ants' nest will contain from 100,000 to 500,000 inhabitants, which will consume every day similar numbers of insects. They also consume plant residues, but, in contradistinction to earthworms, are probably more active in humifying insects than plants. The nature of their diet must be an important factor influencing the composition of the total soil fauna, but little is known about this. Reference has already been made to the action of ants in forming a soil structure, and they are, perhaps, as effective as earthworms in this respect—presumably by excreting humus-clay in the form of soil crumbs. On some poor acid grasslands where earthworms are absent the structure-forming activities of ants which can live under such conditions may be of great significance. The size and frequency of anthills in many places are an indication of the cultivation and soil mixing that some millions of these industrious insects can do. Krausse (1916) estimated that the ant *Formica fusca cinerea* Mayr brought 528 kg. of subsoil to the surface on 100 square metres of sandy soil, or about 20 tons to the acre.

There are a large number of other faunal groups, including slugs, snails, caterpillars, centipedes, millipedes, spiders, flies, beetles, lice, mites, springtails, etc., each of which must play a more or less definite part in maintaining the stable equilibrium of life that prevails in uncultivated soil. What these parts are we do not yet know, but we do know that the total soil fauna comprises a highly efficient machine for converting into soil all organic residues produced by the plants associated with the fauna. Some animals (or insects) eat only fresh plant residues, others require the residues to be partly broken down by micro-organisms before they eat them. Similarly, some bacteria attack fresh plant material while others require it to be passed through an animal body first. In the course of being turned into humus and soil, plant material may have to pass through, or become part of, a whole series of animals and micro-organisms. It is surprising what a perfect complement the natural fauna is to the natural flora. The job is done with remarkable efficiency. I have seen a pine forest in which recently pruned branches covered the ground loosely to a depth of three feet or more ; in an adjoining section similarly pruned five years before, although nothing had been removed, there was scarcely a trace of the prunings to be seen. Everything had been decomposed, and the forester who had lived among such things all his life naturally saw nothing remarkable in them. Yet it is inconceivable that if those tree branches (some several inches thick) had been spread in an open field they would have disappeared so completely. The forest fauna and microflora, however, dealt with them without difficulty.

MICROFAUNA

The numbers of microscopic animals in fertile soil must be very large, but their mass, relative to the mass of the entire soil fauna, is very small—perhaps 1–2 per cent. Nematodes or eelworms are the most numerous group, some species of which, like the sugar-beet eelworm, are parasitic on crops and may cause serious losses, but most nematodes seem to be harmless. Probably they have a special role to play in the humification of plant and animal residues, but what that special role is we do not yet know. The soil contains pores of all shapes and sizes, and it may be taken for granted that some form of life will find in every hole, however small, a suitable home.

The smallest members of the soil fauna are the protozoa, which consist of only a few cells each. They live in the soil water, and when the soil dries out completely they go into an inactive cyst form in which they can exist for long periods. They live mainly on bacteria, and at one time it was thought that by selectively destroying protozoa and thus allowing beneficial bacteria to proliferate freely it would be possible to increase the productivity of soil. This was the explanation given of the benefits of 'partial sterilisation' (p. 95), which was believed to kill protozoa but not bacteria, but recent evidence indicates that the total bacteria-destroying effect of protozoa is small. However, as a result of the interest roused in protozoa by the earlier belief a good deal more is known about them (though not much of immediate practical use) than about most members of the soil microfauna.

CHAPTER VI

THE SOIL POPULATION: II FLORA

MACROFLORA

BY far the most important group of organisms in the whole soil population consists of the higher plants. They are so important that they are liable to be overlooked in discussing the soil population, which is sometimes treated as though the only things that really mattered were micro-organisms and earthworms. The neighbourhood of the roots of the higher plants is where most soil-forming processes occur, and in the absence of roots these processes cease or are greatly weakened. Plant roots profoundly affect the physical properties of soil, compacting it as they grow through it and subsequently, when they die and decompose, providing a source of humus and leaving channels and spaces into or through which air and water may be able to pass. The compacting of a soil richly provided with actively decomposing humus, and the breaking-up of the compacted soil into crumbs by the action of a mat of grass roots, is the only known way in which the ideal soil structure for cultivated crops can be produced. The differing root systems of other plant associations than a grass sward (a forest, for example) will have different effects on soil structure, but all root systems tend to compact soil in the neighbourhood of growing roots, to provide humus in the neighbourhood of dead roots and to leave the soil porous where roots have decomposed—in other words, in varying degrees to granulate the soil. Plant roots also help to hold the soil in place and protect it from water and wind erosion ; without this help there would, in many parts of the world, be no soil at all.

The properties of the soil also influence the development

of the plant-root system.   A root cannot penetrate into a very compacted soil.   There must be at least a small channel into which it can start to grow, and it needs both air and moisture for its growth.   Consequently, roots grow more rapidly and more profusely in open than in compact soils, and they do not penetrate below the depth to which the soil is moistened. Plants, like animals, do not work for the fun of it, and will not make an unnecessary effort to produce roots.   Plants growing in fertile well-moistened soil can get all the nutrients and water they need from a small volume of soil, and produce dense compacted root systems, whereas plants which have to forage farther for their bare requirements produce diffuse root systems.   Tomato plants growing in water cultures have actually been known to wilt from lack of moisture because life was so easy for them that they did not develop sufficient roots to absorb all the water they needed in the period of greatest growth.

The amount of organic material produced by the roots of, say, a grass crop is very considerable, and may amount to 1–2 tons per acre per year.   A cereal crop may produce 5–10 cwt. per acre, which together with stubble represents a not negligible contribution to the organic-matter content of the soil.   Artificial fertilisers, which contain no organic matter, can thus help to maintain the humus content of the soil by producing a bigger crop of roots.   The plot receiving complete fertilisers on the permanent-wheat field at Rothamsted and yielding about 35 bushels of grain per acre contains about 20 per cent more humus than the plot receiving no fertilisers and yielding about 12 bushels per acre.   Much of the root material consists of very fine root hairs, whose diameter is measured in thousandths of an inch, that have an ephemeral existence and are continually dying off, decomposing and being renewed.   The total length of such root hairs that may be produced is enormous.   Pavlychenko (1942) estimated that an oat plant having ample growing space and free from competition produced nearly fifty miles of roots in eighty days, but only about half a mile of roots under ordinary growing conditions.

The continuous dying-off of fine roots which are readily decomposed provides very favourable conditions for microbial development, and the soil in the neighbourhood of plant roots —known as the *rhizosphere*—affords a special kind of habitat for micro-organisms. It is probable also that plant roots excrete inorganic substances, accessory growth factors (hormones, vitamins) and amino acids which stimulate microbial life. At any rate, the rhizosphere is the main seat of micro-organic activity in the soil.

The vegetation, including both the roots and the above-ground portions, is the ultimate source of the organic carbon which is present in all soils in the form of humus. The carbon is obtained from atmospheric carbon dioxide by photosynthesis in the leaves of living plants, and when the plants die their dead residues return to the soil and undergo many processes of decomposition and resynthesis, finally forming humus which is discussed more fully in Chapter VII. Although the intermediate products of decomposition vary with the nature of the plant material and the environment, the final product, humus, appears by our present crude analyses to be much the same whatever the vegetation may have been. Plants and their residues, however, differ among themselves in the amounts of minerals and plant nutrients they contain, and these nutrients become available to another generation of plants as the residues decompose and humify. Consequently, plants which absorb large quantities of nutrients and draw on a soil's fertility reserves also tend to maintain fertility in active circulation if their residues return to the soil. In particular, high nitrogen and lime contents in plant residues promote their active decomposition and a rapid turnover of nutrients in the soil, and it is for this reason that legumes usually make excellent green manures. Plant residues like pine needles which contain little nitrogen, lime or other nutrients tend to decompose slowly and may not provide the soil with sufficient of the

elements of fertility to compensate for what is lost by leaching
by rainwater. Thus plants like pine trees which make very
small demands on soil fertility may actually exhaust the soil,
while the growth of more exacting plants may increase its
fertility.

Plants which require large amounts of nutrients will
naturally grow on soils which can provide the nutrients, and
vice versa. Also, plants which make the biggest demands on
soil fertility will make the biggest contributions to soil fertility
in their residues. An equilibrium level of soil fertility, depending on the inanimate factors of soil formation—climate,
geology, topography, etc.—is thus set up at which the fertility
elements taken from the soil by the growing vegetation is
balanced by the fertility elements restored to the soil in the
dead vegetation. But whereas under so-called natural conditions a balance between income and outgo of soil fertility
is automatically maintained by the whole complex of living
things, under agriculture the balance is unavoidably upset by
the removal of fertility elements in crops. The art and science
of soil management consist in discovering how to maintain or,
better, to enhance fertility by recompensing the soil in various
ways for the economic imbalance resulting from the practice
of agriculture.

Microflora

The most numerous group of the microflora of an ordinary
soil comprises the bacteria of which very many kinds with
apparently different functions are known, and many others
are not known, but certainly exist. Besides bacteria the soil
microflora consists of actinomycetes, moulds (fungi) and algae,
in increasing order of size and decreasing order of numbers.
There are also micro-organisms such as the myxomycetes,
which might be regarded as either animal or vegetable or both.

As already mentioned, the bulk of the bacterial population
is concentrated in the neighbourhood of plant roots—in the
rhizosphere, where the constant sloughing-off of rootlets

provides the bacteria with a rich source of energy. Pockets of decomposable organic matter such as plant residues, manure and compost also form centres of bacterial activity, but humus itself, in the sense of the term as the end product of microbial decomposition, is not a good medium for bacterial development. The majority of soil bacteria thrive best in the absence of marked acidity, and in acid soils the microbial population consists predominantly of fungi which, as a class, are more tolerant than bacteria of acidity, and are unable to decompose organic matter to such an advanced stage of humification. In acid soils fungal matter can sometimes be seen in the form of thread-like or cobweb-like growths (hyphae and mycelia) intertwining the soil particles. They are said to help to break the soil material into aggregates or crumbs, but in this respect they are not nearly so effective as the roots of grass.

It is probable that the soil population contains organisms, active or dormant, capable of decomposing almost every kind of organic compound—otherwise the resistant compounds would have accumulated in places during the long time in which life and soil have existed. Organic substances which do accumulate in special circumstances are peat and coal, but such forbidding materials as carbolic acid and plastics (some of which make excellent nitrogenous fertilisers when decomposed) are dealt with easily by soil bacteria. The soil is also the source of many antibiotics, like penicillin, now being produced to fight microbial diseases. Antibiotics are excreted by a variety of soil micro-organisms, and must serve an important but at present little-understood function in maintaining the balance of microbial life. They probably help to keep down the spread of pathogenic organisms ; indeed, soil is such an excellent medium for microbial development that it is difficult to see, were it not for the antibiotics in the soil, how any bacterial epidemic could ever have been kept within bounds. If there is anything in the belief that organic manuring reduces the incidence of plant diseases it is probably that organic manures increase the microbial population of the soil, and consequently the output of antibiotics lethal to plant pathogens. Several scientists (Winter, 1949) have pointed out that the concentration of micro-organisms in the rhizosphere produces an 'antibiotic buffer' against soil-borne plant diseases.

A great deal of attention is now being paid to the composition of the soil microflora, and especially of the rhizosphere, but there is little we can say now that will not be out of date in a few years' time. There is some evidence that each crop has a specific type of rhizosphere population. If this is so, then one effect of continuous cultivation of one crop would be that the microfloral population would become too specific, and the general microbial equilibrium might be disturbed. On the other hand, it may be possible to arrange crop rotations so that the microflora promoted by one crop will destroy disease organisms of the following crop. Thus Krasilnikov(1949) states that the rhizosphere microflora of lucerne destroys the causative organism of cotton wilt. Glynne and Moore (1949), surveying wheat crops in Britain, have shown the close dependence of soil-borne diseases on previous cropping, and that the risk of trouble can be assessed before sowing. The two most serious yield-reducing fungal diseases, take-all causing stunting and eyespot causing lodging, can be largely controlled by rotating cereals with the other crops on which the disease-producing fungi cannot survive. Controlling these diseases by crop rotation, yields of wheat have been increased in Rothamsted experiments by as much as half a ton an acre without altering fertiliser treatment. Systematic planning of rotations to reduce soil-borne diseases could undoubtedly greatly increase yields. This opens up fascinating possibilities for the future of getting bigger and better crops by rotations planned on microbiological principles, but at present, like so much else in soil science, they are only possibilities.

Bacteria

Although the numbers of bacteria in the rhizosphere are high in comparison with the rest of the soil, the number of species present is relatively small.   Most of the species are non-sporing bacteria active in the decomposition of fresh plant residues, and not in the breakdown of humus (Katznelson, Lochhead and Timonin, 1948).   Owing to the uncertainties and inaccuracies in the methods used for counting bacteria it is difficult to estimate bacterial populations in soil, but figures of the order of 5,000 million [1] per gramme have been obtained for manured soils.

In the following table are shown some counts, in millions per gramme, of bacteria in a ' control ' soil (presumably carrying no crop) and in the rhizosphere of the same soil carrying timothy grass, lucerne or clover (Lochhead, 1948).

Control	Timothy	Lucerne	Clover
133	826	830	2,650

With a population of 1,000 million per gramme of soil the bacteria would weigh about one thousandth part of the mass of the soil, or, say, 1 ton per acre.   Krasilnikov (1944) estimated that the weight of bacterial matter in the top ten inches of a soil carrying three-year-old lucerne to be $2\frac{1}{2}$ tons per acre, in a colder, podzol soil carrying three-year-old clover to be 0·35 ton per acre, and in a similar soil carrying wheat to be only $5\frac{1}{2}$ lb. per acre.   These figures are typical of the very wide variations to be found under different soil and cropping conditions, and the highest figure indicates that a large proportion of soil organic matter (humus) may be derived from the bodies of micro-organisms.   It should be noted, however, that bacterial numbers fluctuate widely from day to day and even from hour to hour within the same soil.

The Nitrogen Cycle

During the process of soil formation inorganic materials of various kinds are transferred from an inorganic environment to an organic biological environment in which they become

---

[1] Or, put into figures and avoirdupois, 150,000,000,000 per ounce

involved in what is usually referred to as the biological cycle or cycle of life. Materials derived from weathering rock or from the atmosphere are absorbed into living plants and bacteria, and thereafter the absorbed elements become part and parcel of the cycle of life ; the organisms die, their residues pass into the bodies of other organisms, undergo decomposition, and the plant-nutrient elements in them ultimately reappear as oxidised inorganic nitrates, phosphates, etc., in the same forms in which they were originally absorbed. The transformations undergone by the different nutrient elements from simple inorganic compounds to highly complex organic compounds and back again to inorganic compounds are known as the nitrogen, phosphorus, etc. cycles.

The cycle of nitrogen is the only one that has been studied in any detail. As stated in Chapter I, there is no nitrogen in the solid raw material—rock—from which mineral soils are produced. The nitrogen that gets into soil all comes out of the atmosphere in which it is present, to the extent of about three-quarters of the total mass, in a chemically inactive form. Small quantities are activated by being oxidised to nitric acid by the action of atmospheric electric discharges, and it is probably in this way that the soil gets its first supply of nitrogen and can then support life. Plants take up most of their nitrogen as nitrate (NO_3) (salt of nitric acid), though some can also take it up as ammonia (NH_3).

With the help of this nitrate a few plants can grow and synthesise carbohydrates and proteins with the aid of sunlight. When the plants die, these organic compounds provide food for a multifarious population of animals, insects and micro-organisms which decompose the plant materials, including the complex nitrogenous proteins. Numerous kinds of bacteria take part in the decomposition of proteins, and numerous intermediate products are formed in the process, but the final products are the simple compounds ammonia (NH_3), carbon dioxide (CO_2) and water (H_2O).

The powerful smell that emanates from a decomposing

manure heap is evidence of the active production of ammonia by bacteria. The ammonia may be regarded as coming from the surplus nitrogen in the plant residues that is not needed by soil micro-organisms in building up their own body protein. The ammonia is available for further plant nutrition, the body protein remains unavailable until it, in its turn, is decomposed and partly ammonified. Ammonia is strongly absorbed by soil particles, and, although its salts are quite soluble in water, it can be stored by the soil when it is not wanted by plants without danger of being washed out by rain.

In a poorly aerated or a very acid soil there may be some accumulation of ammonia, but in a well-aerated and not too acid soil (i.e. most cultivated soils) ammonia is further transformed first into nitrite (NO_2) by the bacterium *Nitrosomonas*, and finally into nitrate (NO_3) by the bacterium *Nitrobacter*. There is little doubt that the microbiological oxidation of ammonia to nitrate does go through the nitrite stage, although nitrite is rarely detectable in soil—which is just as well as it is harmful to plants. The second stage—NO_2 to NO_3—is much more rapid than the first stage—NH_3 to NO_2. Unlike ammonia, nitrates are not absorbed by soil. They are very soluble in water, and if there are no plants to utilise them as soon as they are produced they are liable to be washed out of the soil and lost to the nitrogen cycle.

Figures for the total-nitrogen, ammonia and nitrate contents of soil naturally vary greatly, but those given by Russell (1932) may be regarded as representative of the proportions in which these three forms occur in arable soil :

Total nitrogen	0·15 per cent
Ammonia nitrogen	0·0002 per cent
Nitrate nitrogen	0·0002–0·002 per cent

Thus at any time only a quite insignificant fraction of the total nitrogen in soil is present in available form. The above figure for ammonia nitrogen, for example, represents about 4 lb. per 9-inch acre—as much as is contained in 20 lb. of sulphate of ammonia—but there is nearly 1,000 times as much nitrogen as this stored up in the soil humus. What is important for plant nutrition, therefore, is not the total quantity of nitrogen in the soil, but the rate at which the nitrogen is made available by further decomposition of organic matter. The rate of decomposition increases with the rise in temperature during the summer, so that an increased supply of humus nitrogen becomes available to growing plants when they need it most, but the rate of supply cannot be greatly accelerated by anything the farmer can do, and to get really high yields it is often necessary, even on highly manured soils, to apply top-dressings of ammonia or nitrate nitrogen.

The Carbon-nitrogen Ratio

We have already said that the ammonia made available during the decomposition of plant material is the surplus not required by soil animals and micro-organisms for making their own body material. Besides nitrogen they also require carbon, in the form of carbohydrates, both for body building and as their main source of energy. The supply of carbon largely determines the size of the soil population. When it is high, relative to the supply of nitrogen (wide carbon-nitrogen ratio), the organisms will multiply until their further increase is stopped by shortage of nitrogen. In these circumstances all the nitrogen will be used for body building, and little or no surplus ammonia available for plants will be produced. There will, however, be a surplus of carbon which the organisms will dispose of in respiration as carbon dioxide which, being a gas, will diffuse out of the soil altogether. Carbon will be lost from the soil system, but not nitrogen, and the initially wide ratio of carbon to nitrogen will be narrowed.

Most plant materials contain a high proportion of carbon to nitrogen. Materials such as straw, leaves and wood contain 50–60 per cent of carbon and 1–3 per cent of nitrogen—that is, they have carbon-nitrogen ratios of 60–20. When they are decomposed in the compost heap or in the soil some of the carbon and all the nitrogen are utilised for body building, and the rest of the carbon is dissipated as CO_2. The carbon-nitrogen ratio falls, and when it reaches a value of about 25, nitrogen begins to be eliminated as ammonia as well as carbon as carbon dioxide. It is at this point that the nitrogen in the decomposing compost heap becomes available as a plant nutrient. Further decomposition still further narrows the carbon-nitrogen ratio until, when the material is completely humified, as in soil, the carbon-nitrogen ratio reaches a constant value of about 10.

The addition to soil of unrotted manure or plant material like straw frequently causes an actual decline in crop yields. This is because the manure or straw has too wide a carbon-nitrogen ratio, which means that the micro-organisms have too much energy material (carbon) in relation to body-building material (nitrogen). Consequently, in order to get more body-building material the soil micro-organisms utilise and lock up not only the nitrogen in the manure, but also the nitrogen in the soil that is available to plants, and crops suffer from lack of nitrogen in spite of the fact that some nitrogen was added in the manure. This is an example of bacteria competing with plants for nutrients. As the manure decomposes further, perhaps in the second year after application, its carbon-nitrogen ratio will narrow, and the nitrogen of which the bacteria robbed the soil will again become available to plants.

It will thus be seen that bacterial activity in soil is not necessarily a good thing—except for the bacteria. We should certainly not identify a high level of bacterial numbers with a high level of agricultural fertility, or *vice versa*. Manuring a soil with starch, which is pure carbohydrate free of nitrogen, would be grand for the bacteria, but very poor for any plants competing with the bacteria for nutrients in the soil. It is believed that the reason why a farmer does not usually recover in his crops more than half the nitrogen he applies in fertilisers is that the rest is taken by competing bacteria. Nevertheless, soil bacteria do the farmer more good than harm ; at least, he does not yet know nearly enough about the soil to do without them !

Nitrogen Fixation

All the nitrogen in soil, except that introduced in fertilisers, is derived from the air above the soil. Mention has already been made of one way in which the chemically inert atmospheric nitrogen is made available to plants by being oxidised to nitric acid and washed into the soil in thunderstorms. This form of 'nitrogen fixation,' as it is called, is, however, almost negligible in a soil's economy, as the amount of nitrogen fixed would not cover normal nitrogen losses by leaching and other causes. The main way in which atmospheric nitrogen gets absorbed into the nitrogen cycle of the soil is by biological fixation by soil micro-organisms, especially bacteria.

There are two classes of nitrogen-fixing bacteria—symbiotic and non-symbiotic. The symbiotic bacteria do nearly all the nitrogen fixing in British soils, indeed it is not known whether non-symbiotic bacteria are effective at all.

Symbiotic nitrogen-fixing bacteria are found on the roots of leguminous plants of all kinds. The legumes cannot thrive without the bacteria, except in very rich soils, or the bacteria without the legumes. The plants provide the carbohydrates (energy material) and the bacteria the nitrogen (body-building material) for the symbiosis. The bacteria produce nodules which can be seen on the roots of any healthy clover, lucerne, pea, bean, etc. plant (Plate 6). They will fix sufficient nitrogen from the air to cover all the needs of their host plant, but the more nitrogen there is already available in the soil the less they will fix. When a legume crop is grown and then ploughed in as green manure all the fixed nitrogen will be added to the soil's supply. A good crop of legumes will fix about 100 lb. of nitrogen per acre, equivalent to the nitrogen in 5 cwt. of sulphate of ammonia or 7 tons of average farmyard manure. It has also been estimated in one of those fanciful estimates that are probably quite correct, but quite incapable of verification, that nitrogen-fixing bacteria supply the soils of the world every year with far more nitrogen than they get from all other sources combined.

The value of leguminous crops for increasing soil fertility has been known since Roman times, but it was not until 1886 that Hellriegel and Wilfarth showed that nitrogen additional to that already in the soil was taken up by legumes, and that bacteria in the root nodules fixed the nitrogen from the air. Beijerinck (1888) isolated the bacterium and named it *Bacillus radicicola* ; it is now known as *Rhizobium*.

There are many strains of *Rhizobium* any one of which will usually live symbiotically with or ' infect ' only certain specific groups of legume, such as the clovers. A legume will not grow well in a soil that does not contain the appropriate bacterium to complete the symbiosis. Native legumes or those which have long been grown in a country are almost certain to find the required symbionts in the soil, but an exotic species may be difficult to establish because its symbiont is absent. The cultivation of lucerne in Britain was for long confined to a few places in East Anglia (mainly in Essex) into which it had been introduced from Holland. It was impossible to get a crop elsewhere until methods were developed about thirty years ago for ' inoculating ' the soil or, better, lucerne seed with the lucerne strain of *Rhizobium*. Now lucerne can be grown in most parts of the country provided inoculated seed is used at first and the soil conditions are suitable for the survival of the bacterium, in which case it will survive for many years even in the absence of the legume host plant.

Recent studies have shown that there are numerous strains of a legume nodule bacterium that may vary greatly in the efficiency with which they fix nitrogen. Several strains may be present in a soil at the same time, all of which may produce nodules on a legume growing in the soil, but each nodule contains only one strain. Ineffective strains of white-clover nodule bacteria are very prevalent in the hill districts of Britain,

and their presence may be a cause of the difficulties sometimes
encountered in establishing clover there. There seems to be
a tendency sometimes for effective strains to evolve into less
effective strains, but the reverse trend has not been observed.
More knowledge of the microbial ecology of soil might show
how to treat soils so as to encourage the development of
effective strains, but that is still only a future possibility.

Nodulation of the ordinary farm legumes will not take
place unless the soil is well supplied with lime and phosphate.
The two rather unusual elements molybdenum and vanadium
are also required for the effective development of nodule
bacteria, but the amounts required are infinitesimal, and are
nearly always present in the soil.

The non-symbiotic nitrogen-fixing bacteria are not
dependent for their activity on the presence of a living host
plant ; they get their nitrogen from the air, and their carbon
from non-living organic matter in the soil. Two types are
known—*Azotobacter*, which is aerobic, i.e. thrives in well-
aerated soil, and *Clostridium*, which is anaerobic, and does
best in the absence of free air, deriving its oxygen from soil
organic matter. It is doubtful whether either of these bacteria
plays a significant part in the nitrogen economy of agricultural
soils, though *Clostridium*, which tolerates more acid conditions
than *Azotobacter*, may do so in forest soils. *Azotobacter* will not
fix nitrogen at pH values below 6, and there are many agricultural soils more acid than that. Indeed, although *Azotobacter* can be made to fix nitrogen quite efficiently when
growing on sugar in a laboratory, it has never been actually
proved that it fixes much nitrogen in soil. Where it does
occur it is usually present in small numbers amounting to
100–1,000 per gramme of soil. Lipman and Conybeare (1936)
estimated that non-symbiotic nitrogen fixation accounted for
about 6 per cent of the nitrogen added to the soils of the
United States in a year, and symbiotic fixation for about
88 per cent.

Following the successful inoculation of soils with the

nodule bacteria of lucerne and other economically important legumes, attempts have been made to increase the nitrogen content of soil by inoculation with *Azotobacter*. All such attempts made on the western side of the Iron Curtain have failed ; but most of those made on the eastern side, of which the results have been published, seem to have succeeded. Whatever may be the cause of this discrepancy, the line of demarcation is quite sharp, and those farming on the western side should not rely on *Azotobacter* to increase the fertility of their soils.

The application of energy-rich organic material, such as sugar, to soil may result, by causing nitrogen-fixing organisms to multiply, in increasing the nitrogen content of the soil, although at first all the fixed nitrogen as well as other nitrogen already present in the soil will be locked up in the body substance of the organisms, and later be released for plant nutrition. Dhar and Mukerji (1936) in India claim that infertile alkali soils can be made fertile by treating them with molasses, which causes a great multiplication of nitrogen-fixing and other micro-organisms which decompose the molasses and at the same time increase not only the nitrogen content of the soil, but also the formation of acids which neutralise the harmful alkalinity. They found that more nitrogen was fixed in the light than in the dark, and concluded that some nitrogen was fixed not by micro-organisms, but by a purely chemical action taking place between the air and the soil, that was activated by light. They thought that this photochemical nitrogen fixation was a major source of nitrogen in tropical soils. However that may be, there is no evidence that photochemical nitrogen fixation takes place to any extent in temperate soils.

On the surface of flooded rice fields a film of blue-green algae forms that is able to fix nitrogen from the air. These nitrogen-fixing algae are of great economic importance in maintaining the fertility of rice fields in tropical countries in many of which nitrogen fertilisers have never been used, but although similar species are known in temperate countries they are not prevalent enough to have any influence on nitrogen economy.

Losses of Nitrogen from the Nitrogen Cycle

Nitrogen in the form of plant residues, soil organisms, humus or other organic material is mostly insoluble in water, and is therefore not removed from the soil by percolating rainwater. If the nitrogen takes an active part in the nitrogen cycle it will ultimately be converted by microbial action into plant-available ammonia and nitrate in both of which forms it is very soluble in water. Ammonia, however, is absorbed by soil and thus protected from being washed out. Nitrate, on the other hand, is not only extremely soluble, but is not absorbed at all. In a moist climate like Britain's, as soon as any nitrate is formed it is either taken up by plants or, if plants are not there or are not growing, it is liable to be washed right out of the soil and lost to the nitrogen cycle. It will ultimately find its way into rivers and the sea, where it will be absorbed by some aquatic organism and take part in another nitrogen cycle. In Britain losses of 20–30 lb. of nitrogen (more in wet areas) per acre per annum can be expected from bare soil, but much less from cropped or covered soil where most of the nitrate formed will be taken up by plants.

It is because these losses are largely unavoidable, both under agricultural and 'natural' conditions, that microbial nitrogen fixation plays such a vital rôle in nitrogen economy. Under agriculture there is a further loss of soil nitrogen, again of the order of 20–40 lb. per acre per annum, in crops and animals sold off the farm, and under certain conditions there is a further loss, the magnitude of which is unknown but is probably small, caused by reduction of nitrate by micro-organisms to gaseous nitrogen which escapes back into the air. The process, known as denitrification, is most marked under conditions of high nitrate and low air supply, for example in peaty soils which are alternately well aerated (when nitrification occurs) and waterlogged (when denitrification occurs).

Fungi

Soil fungi are larger organisms than bacteria. They can be as large as giant puffballs or of microscopic size. Their numbers may vary from thousands to a million or more per gramme of soil, and the total weight of fungi is probably of the same order as of bacteria. As a class of organism fungi are more catholic in their tastes than bacteria, many of which perform one specific function in the soil, such as nitrogen fixation, cellulose decomposition, ammonification, nitrification, decomposition of sulphur or iron compounds, etc. Fungi might be regarded as the scavengers which will decompose almost anything that bacteria cannot tackle and much that they can, and, having done so, many of them serve as food for bacteria. They are the chief decomposers of lignin, the fibrous, bacteria-resistant compound which is the main constituent of wood and straw, and makes these materials difficult to decompose in compost heaps. Fungi are often the predominant kind of micro-organism in forest soils, where they serve the purpose of disposing of the great mass of lignified material which falls on to the soil as dead tree branches and twigs. Since humic acid, the part of humus extracted from soil by alkali, is composed largely of substances chemically similar to and obviously derived from lignin, it seems as though fungi were among the chief humus-forming micro-organisms of the soil.

It has been suggested by a Russian, Geltser, that humus is the end product of the autolysis (i.e. the self-decomposition by enzymes) of the dead bodies of bacteria which have consumed the bodies of fungi which have consumed the residues of plants. This is an attractive hypothesis, as it brings a whole series of biological agents into humus formation in an orderly way, but it has not yet received much supporting evidence.

Mycorrhizas

Numerous fungi inhabiting the soil cause diseases to plant roots (see *Plant Diseases*, by F. C. Bawden in this series), but the majority are harmless and some are beneficial to crops. Some fungi live in close association with plant roots, a special association of this kind being known as *mycorrhiza* (from the Greek *mucor*, mould, and *rhiza*, root). Mycorrhizas can take many forms, some beneficial, some harmful to the host plants, and some with no obvious effect on them. In the beneficial forms the fungal hyphae seem to live in both the soil and the root, getting much of their organic nutriment (carbon) at the expense of the plant, and repaying the plant by providing it with nitrogen and minerals from the soil that enable the plant to photosynthesise far more organic matter than it would otherwise have done or than the mycorrhizal fungus consumes. Thus both plant and fungus benefit from the association—especially the fungus, which often cannot live without the plant root, though the plant can live without the fungus.

Mycorrhizas have been found on a very large variety of plants, including agricultural crops, but so far no benefit from the mycorrhizal association to any agricultural crop has been observed. The chief beneficiaries are coniferous trees, especially when growing in soils of rather low fertility. The mycorrhizas can be seen as short branched excrescences on the smaller roots (Plate 7). The excrescences are covered with fungus mycelium, and the fungus hyphae penetrate into the cortical cells of the root. In normal fertile soils all trees will have mycorrhizas, and it is therefore impossible to tell whether the mycorrhizas benefit the trees or not. In very infertile soils practically no trees will have mycorrhizas, since the symbiont fungus needs a certain minimum of plant nutrients, particularly nitrogen and phosphorus, and in such soils very striking results have been obtained by inoculating soil in the neighbourhood of the roots of young trees with the appropriate fungus (Rayner, 1943). Starved and stunted

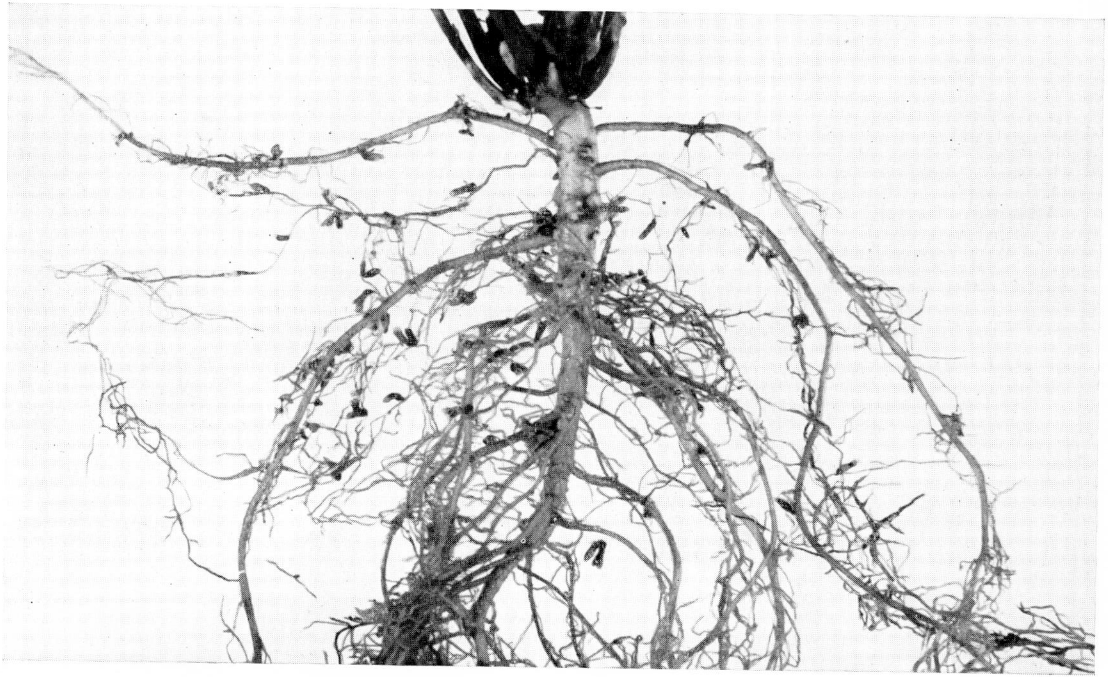

Plate 6   Bacterial nodules on lucerne roots

Photo : V. Stansfield

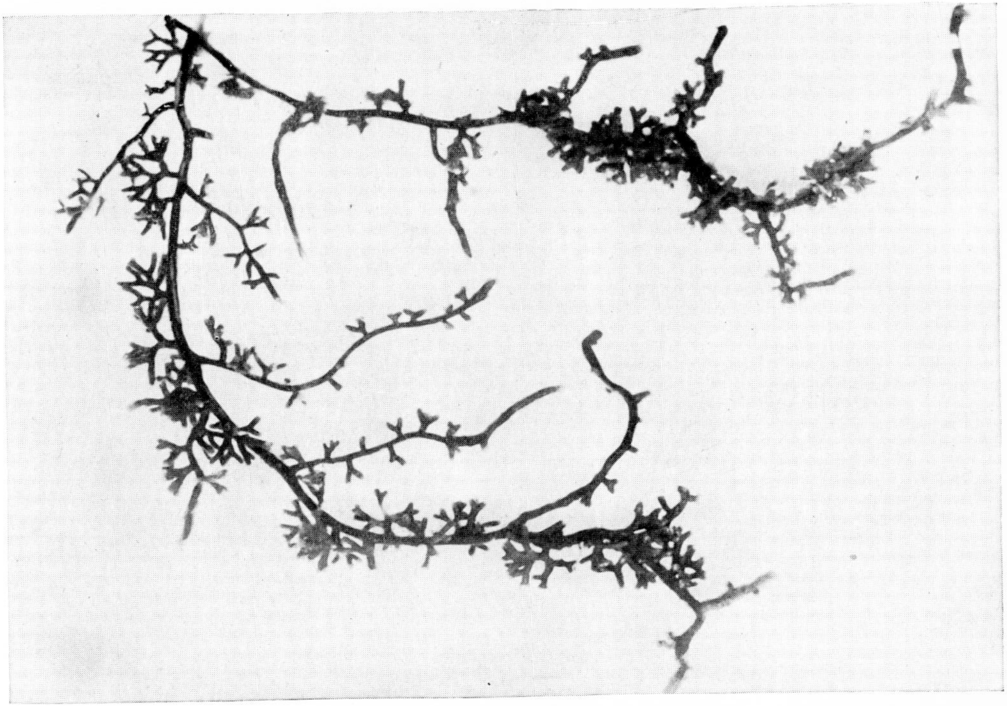

Plate 7   Scots Pine roots showing mycorrhizal development

By courtesy of Faber & Faber Ltd

young trees rapidly assume a healthy green appearance and grow well. The effect is very similar to what might be obtained from a balanced application of fertilisers. Analyses of mycorrhizal and non-mycorrhizal pine seedlings grown under comparable conditions have shown that the former contain much more nitrogen, phosphorus and potassium than the latter (Melin, 1948). Somehow, mycorrhizal roots seem able to take up considerably more plant nutrients from the soil than non-mycorrhizal roots. One, but by no means the whole, explanation of this is that a mycorrhizal root has a bigger absorbing surface.

Conifer mycorrhizas, in which the fungus is partly inside and partly outside the plant root, are ' ectotrophic.' There are also, and more commonly, ' endotrophic ' mycorrhizas in which the fungus lives entirely within the root and there is no visible sign of the mycorrhizal association. Endotrophic mycorrhizas are the kind which is found in agricultural plants, but their significance, if any, is unknown. There is no definite evidence that they have any effect on crop growth or yield, though there is a little evidence that they are more prevalent on nutrient-deficient than on nutrient-rich soils.

Partial Sterilisation of Soil

There is a limited connection between the capacity of a soil to produce crops (what is commonly known as its fertility) and its capacity to produce micro-organisms. As might be expected, conditions which favour crop production usually also favour microbe production, but it must be remembered that microbes compete with crops for the nutrient supply of the soil. We have already seen that the weight of microbial matter in a soil may be several tons per acre—the same order of magnitude as the total weight of a cereal crop, and we have seen that microbes lock up, temporarily at least, large quantities of plant nutrients in their bodies. This suggests that if we could get rid of soil micro-organisms and at the same time do their job of ' cooking ' the food of plants we

might double crop yields because we should double the supply
of plant food. Artificial fertilisers now supply plant food ready
' cooked,' and have greatly diminished the rôle of micro-
organisms in the plant-food economy of agricultural soils ;
nevertheless the decomposition of organic matter by micro-
organisms remains important, owing to its effect in developing
and maintaining the structure of the soil. A soil that was
completely deprived and kept sterile of micro-organisms
would be unlikely to retain its crop-producing capacity for
long, even if it were amply supplied with plant nutrients,
because its structure would deteriorate.

By what is known as ' partial sterilisation,' the micro-
population of the soil can be temporarily reduced and at the
same time its productivity, as expressed in plant growth,
greatly increased. Soil is treated either by steaming, which
kills off all but the most heat-resistant micro-organisms, or
with volatile antiseptics such as formaldehyde or carbon
disulphide, which also kill off most of the micro-organisms
and then disappear by evaporation. Partial sterilisation is an
expensive method of increasing productivity, and, as its effects
only last for one growing season in the open, it is not used
except in glasshouse culture, where the extra yield will pay
the cost of the treatment, or in special cases outdoors like
forest nurseries where the value of the crop per acre is very
high.

How and why partial sterilisation works in this way is not
known. In glasshouse (tomato) culture where sterilisation is
now standard practice, one of the main effects is undoubtedly
to destroy disease-producing organisms, and the treatment is
sometimes known as ' disinfection.' In nursery practice a main
effect is to destroy the seeds of weeds which otherwise tend
to suppress the tree seedlings. But these effects do not explain
all the phenomena associated with partial sterilisation that
include delayed germination of seeds sown, followed by
vigorous and luxuriant growth which may continue throughout the growing season or peter out.

It was formerly thought that partial sterilisation destroyed organisms which competed with plants for plant nutrients, so that when they were killed a larger amount of nutrients was available for plants. For several reasons, however, this explanation is not satisfactory, and another, put forward later by Russell and Hutchinson (1909), that partial sterilisation selectively destroys protozoa which prey on bacteria and thus allow the latter to multiply, has been discarded as untenable because the numbers of protozoa normally present in a soil are very small relative to the numbers of bacteria. The most striking direct effect on the soil of partial sterilisation that has been observed is an initial great increase in ammonia production, followed later by an increase in nitrate production. As a result plants, when once they have got over the retardation of their germination, have at their disposal abundance of available nitrogen. Sterilisation also produces an increase in the quantities of soluble mineral nutrients (phosphates, potash, lime) in the soil, though nobody knows why. It has been pointed out that many of the effects of partial sterilisation are similar to those of a bare fallow, and the immediate, though perhaps not the ultimate, causes may also be similar, namely, destruction of plant-disease organisms and weed seeds and an extra accumulation in the soil of available plant nutrients. Sterilisation, either by steam or volatile antiseptics, has been shown to have certain quite definite effects on a soil's physical properties, but they are not all beneficial to plant growth. The most noticeable microbiological changes are a temporary cessation of nitrification (which allows ammonia to accumulate) and the almost complete disappearance of the fungal population. The wholesale destruction of fungi has only been observed in the last few years, and its significance is not known.

Algae

Soil algae are microscopic, chlorophyll-containing organisms which, owing to their chlorophyll content, are capable of synthesising organic matter from the carbon dioxide of the air, thus ' fixing' atmospheric carbon in the same way as *Azotobacter* and *Rhizobium* fix atmospheric nitrogen. The amount of carbon so fixed must be minute compared with the amount fixed by plants, and as far as is known soil algae do not play a significant part in the economy of agricultural soils (except the nitrogen-fixing algae of rice fields referred to on p. 91). Algae may, however, have a greater significance in the very earliest stages of soil formation. They are sometimes the first chlorophyll-containing organisms to inhabit bare rock. With the carbon dioxide they collect from the atmosphere they produce the first traces of soil organic matter which provides the pabulum for the first bacterial inhabitants which prepare the ground for the first of the higher plants. They are also believed to be powerful agents in the decomposition of the primary minerals of rocks and the synthesis of clay minerals.

CHAPTER VII

HUMUS

WHEN all the plants and animals and micro-organisms have eaten and digested each other until they can get nothing more out of each other, what is left is known as humus. The term humus is usually taken to include any plant, animal or microbial susbstance that has lost its original structure and been reduced to an amorphous colloidal state. By then it has passed through so many transformations that it is not plant, animal or microbial in origin, but all these. Humus is ordinarily dark coloured and gives soils their characteristic brown, grey or black colour, but humus is not necessarily dark, for in the tropics there are bright red and yellow soils containing appreciable quantities of humus. Humus is not a definite substance ; it is in a continual state of flux, disappearing by decomposition and oxidation and being reformed from the residues and excrements of organisms. Nevertheless in any soil carrying permanent vegetation the nature of the humus does not change very much with time. The humus is the end-product of all the soil's organic activities, and so long as these remain unchanged the humus will also remain unchanged in nature though not unchanged in substance.

Almost certainly the humus of one type of soil differs from that of another, but it is not easy to state what the difference, chemically, is. Sometimes the difference between extreme types is obvious—as that between the peaty raw humus of heathland and the crumby earthy humus of a well-manured garden. But when we isolate the humus from the non-humus parts we find little chemical difference between the heathland and garden products. One reason for this is that the very act of isolating humus, by extracting soil with alkali and precipitating ' humic acid ' with an acid, changes the nature of the original material, but all humic acids prepared from soils, peats, composts, etc. have much the same chemical composition and properties. Our knowledge of these properties is the main foundation for the current theories on how humus is formed.

CHEMISTRY

When soil or peat is treated with caustic-soda solution a dark-coloured liquid is obtained from which humic acid is precipitated by adding acid. Humic acid, so prepared, looks when dry rather like coal. It is completely insoluble in water, but dissolves readily in caustic soda forming soluble sodium humate. Both the sodium and potassium salts of humic acid are soluble, which is the reason why humus, in the form of sodium or potassium humate, is washed out of the surface of certain alkaline soils containing much soda or potash. Under very acid conditions, such as give rise to the formation of peat, water-soluble organic acids, known collectively as fulvic acid, are formed. They produce the dark colour that tinges water draining from a peat moor, and they are probably the form in which humus moves through a podzol, removing the iron and aluminium from the surface mineral soil and depositing them lower down (Plate 1a).

The chemistry of humus is extremely difficult, and very little progress has been made in elucidating it despite the labours of hundreds of first-class organic chemists. It is impossible to obtain a 'pure' humus product of consistent composition and properties. Humus is nearly always studied chemically in the form of humic acid, which is the purest condition in which it can be obtained nearly free of mineral matter and unhumified organic matter. Up to three-quarters of the humus of an ordinary soil can be extracted with alkali, and may therefore be assumed to be present in the form of humic acid or its salts (humates), though there is also an alkali-insoluble fraction, known as humin, that seems to be resistant to decomposition and is assumed to be the least active part of humus. The percentage chemical composition of humic acid is approximately : carbon 56, hydrogen 5, nitrogen 5, oxygen 34. There is, besides, a variable amount of mineral matter which contains phosphorus, iron, silicon, and other elements in smaller proportions.

Humic acid behaves chemically like the substance known as lignin, which can be obtained from many plant materials by treatment with dilute alkaline solutions. Lignin is the substance that gives wood, straw and leaves their rigidity, and is one of the most resistant of plant constituents to microbial attack, besides being useless as animal food. Consequently it accumulates in the soil relatively to other plant constituents like cellulose, starch and protein which are decomposed by one or more soil micro-organisms. Thus humic acid—what is left when soil organisms, large and small, have finished with plant and animal residues—consists partly of lignin-like substances derived directly from the breakdown of plant residues and partly of new synthesised substances which have been built into the bodies of the soil fauna and micro-organisms in consuming and digesting the non-lignin constituents of plant residues. These include compounds related to carbohydrates and proteins. Humus, or humic acid, is a product of both biological decomposition and biological synthesis. Little is known of the nature of the synthesised portions of humus, but they always contain nitrogen, phosphorus and iron. The nitrogen and phosphorus compounds undergo gradual oxidation, at the end of which they appear in forms available for plant nutrition. Humus is, indeed, the main source of both nutrients under ' natural ' conditions of plant growth. The rôle of the iron in humus in not clear, but it may be concerned in chemical, as distinct from biological, oxidation.

Decomposition

One of the unsolved mysteries of humus is how and why it disappears from soil at all. It is the end-product of biological activity, and although it has been stated that actinomycetes thrive on humus most other soil organisms leave it alone. Yet in an ordinary uncultivated soil like a pasture soil it disappears completely as quickly as it is formed ; there is no appreciable increase in the humus content over a long period of years. The resistant lignins of dead plants disappear as quickly as do the carbohydrates and proteins which microbes, like animals, find more to their taste. On the other hand, in a cultivated soil to which no organic matter is added the humus content falls quite rapidly at first, then more slowly. This suggests an easily decomposable fraction (carbohydrate and protein ?) of humus that disappears slowly, but it does not explain why the stable fraction does not accumulate steadily when new organic matter is continually being added to soil—as under forest or pasture. As far as is known, both the quantity and quality of humus in soil under permanent vegetation remain constant. Under cultivation the quantity falls, but the quality does not seem to change much, though it must be remembered that our methods of assessing humus quality are crude.

The indications are that in mineral soils humus disappears as a whole, and since no micro-organisms or groups of micro-organisms are known that will effect its complete oxidation that process must be at least partly non-biological, i.e. physico-chemical. It may be that the iron present in all soil humus acts in some way as an oxidising catalyst. It is significant that humus does tend to accumulate in non-mineral systems like peat, compost and manure, but when these materials are incorporated with soil the humus goes completely. The actual incorporation of humus with soil is effected mainly by the soil fauna ingesting and excreting both plant residues and clay (Chapter V), and the product, a combination of colloidal humus and colloidal clay known as the clay-humus complex, is physico-chemically much more active than either the clay or the humus alone. It is probably only when in the form of this complex that humus undergoes complete oxidation.

Nothing is known about this oxidation of humus, or even if it really happens, but something like it must happen to explain why the microbially resistant humus products which do accumulate in purely organic formations like peat do not go on accumulating in organo-mineral formations like soil. Each soil seems to have a maximum capacity for humus, and when once that capacity has been reached the humus content does not rise appreciably however much (within limits) organic matter is added.

Humification

At any rate, the process of humus oxidation is quite different from the process of humus formation or humification ; it is not just the final stage of the decomposition of plant residues. Humus itself is largely a synthetic substance, synthesised by animals and micro-organisms in the passage of plant residues into and through their bodies. Except for the part that comes from lignin, humus has very little chemical affinity with products synthesised by plants. Humification in composts is usually accelerated by lime because a neutral or alkaline reaction encourages bacteria which decompose plant residues. In soils, on the other hand, lime tends to increase the capacity to retain humus unoxidised. The mineral soils with the highest humus content are the lime-rich chernozems of steppes, and in Britain similar high humus contents are found in the black soils overlying chalk. This effect of lime, however, is probably an effect on the physical properties of clay colloids, enabling them more readily to form a clay-humus complex rather than an inhibiting effect on humus oxidation. It is known that the calcite secreted by earthworms (p. 70) plays an essential part in the formation of the clay-humus complex excreted by the worms at their other end. We cannot with our present knowledge explain further the humus-conserving effect of lime, but it does throw into contrast the difference between the process of plant decomposition leading to the formation of humus, and the process of humus decomposition.

(1,030)

Nitrogen

The element in humus that is of most interest to the farmer is nitrogen. It is not strictly correct to say that humus is the sole source of nitrogen for plants in uncultivated soils because plants can also obtain nitrogen, without its passing through the humus state, from nitrogen-fixing organisms and from decomposing plant residues before they are actually humified. Nevertheless, about 98 per cent of the nitrogen in most soils is in organic form, and the other 2 per cent is present as ammonia or nitrate, most of which has been released by bacterial action from organic combination.

There is a good deal of evidence that much of the humus nitrogen is in the form of protein, although no protein has actually been isolated from soil. There are grounds for believing that the main part of humus—or, more correctly, humic acid—is a combination of lignin and protein [1] with bits added on to explain properties not attributable to a ligno-protein complex. A very significant property of humus protein is that, quite unlike all animal and vegetable proteins, it is very resistant to microbiological attack. It has been found that when lignin and protein are artificially combined the protein becomes very resistant to biological decomposition (Waksman and Iyer, 1932, 1933), and a somewhat similar protection against protein decomposition is given by the presence of certain clay minerals, especially of the montmorillonite group (Ensminger and Gieseking, 1942). Whatever the reason for the stability of humus protein, the stability itself is a vital factor in soil economy, ensuring that nitrogen stored in humus becomes available to plants as ammonia and nitrate in a steady stream throughout the growing season and not, as would otherwise happen, all at once when things get busy in spring.

[1] The ' ligno-protein complex.' Soil scientists are very fond of complexes ; they make things seem less complex !

Phosphorus

The phosphorus compounds in humus are likewise only gradually made available to plants by the decomposition of the humus, but, unlike nitrogen which is almost entirely in organic combination in soil, phosphorus may be present to the extent of half or more of its total content in inorganic mineral form. In uncultivated soils plants are not solely dependent for their phosphorus nutrition, as they are for their nitrogen nutrition, on the decomposition of organic matter. Nevertheless, there is some evidence that the plants normally get a large proportion of their phosphorus from organic sources, that is, by the oxidation of humus phosphorus compounds to phosphate. After a few days or weeks all soluble phosphates become ' fixed ' (unavailable to plants) by interaction with soil minerals, but when there is a continuous slow oxidation of organic phosphorus to phosphate there is a period between the completion of the oxidation and the subsequent fixation when the phosphate can be absorbed by plants. There is also evidence (Garman, 1948) that organic soil phosphorus is about as available to plants as is inorganic phosphorus. Some of the organic phosphorus seems to be in the form of nucleic acids and their derivatives, some of which are utilisable by plants.

Potassium

The other major plant nutrient, potassium, does not occur in humus except as the potassium salt of humic acid—potassium humate. In this form potassium is probably as available to plants as it is in the exchangeable form adsorbed on clay, and there is, in fact, no way of distinguishing between humate potassium and exchangeable potassium. Whatever part humus plays in the potassium nutrition of plants can be performed equally well by the mineral colloids. The same cannot be said of nitrogen and phosphorus, for which humus acts as a kind of self-adjusting control, releasing the nutrients rapidly when

conditions are favourable for plant growth and slowly when
they are not.

Trace Elements

The trace elements, including manganese, copper, zinc and
boron, are known or can be assumed to be present in very
small amounts in soil organic matter.   Being essential to
plants they will occur in plant residues, and will become available to plants when the residues decompose.   Trace-element
deficiencies can often be cured by applying farmyard manure.
Humus, according to Waksman (1936), offers a very favourable source of available iron for both plants and micro-
organisms.   Soil organic matter, including not only humus
but plant and animal residues in all stages of transformation
into humus, is, indeed, a very fine food for plants, containing
everything a plant needs and supplying it to the plant when
it needs it.   But it is economically inefficient as a source of
plant food.   The stable, if low-geared, economy of a meadow
or forest illustrates how reliable and unfailing such plant feeding is, but modern agriculture requires very much greater
economic efficiency than anything wild Nature achieves,
and the use of organic manures, except to improve a soil's
physical condition, is falling into desuetude in industrialised
countries.

Hormones

The organic matter of the soil contains small quantities of
growth-promoting substances or ' plant hormones,' which have
a powerful effect on plant growth, and perhaps play an analogous rôle to vitamins in animal nutrition.   That plants can be
grown perfectly well in water cultures shows, however, that
organic growth-promoting substances are not indispensable to
plant growth, but it is probable that they do help seedlings
to root in soil.   Some proprietary articles similar to natural
growth-promoting substances are now sold to encourage active
root growth in seedlings and cuttings.   The selective ' hor-

mone' weed-killers are also growth-promoting substances whose action depends on their promoting growth so vigorously that they completely upset the metabolism of some plants, mainly dicotelydons, which die.

PHYSICAL EFFECTS

Humus has profound effects on a soil's physical condition. It is the substance which gives a soil structure. The beneficial effects of humus are most noticeable in sandy soils and clayey soils. The former have pores which are too large to retain water, the latter pores which are too small. Humus binds the particles of a sand together, and forces those of a clay apart, giving both soils greater water-holding capacity. Humus is a colloid with large swelling capacity which, together with its ability to combine with clay minerals as already described, enables it to act as the glue or ' tissue ' binding the separate mineral particles into porous crumbs.

Bacterial Gums

The nature of the glueing material has been the subject of much research, and it is now believed that it may be gums synthesised and excreted by bacteria. These bacterial gums consist largely of ' uronic polysaccharides ' formed from various sugars like glucose, mannose, xylose, etc., and sugar derivatives of the nature of glucuronic and galacturonic acids. A well-known non-bacterial substance of this type is pectin. The apparent fact that soil organic matter is not a structure-former unless it is actively decomposing lends support to the belief that the glueing material is of bacterial, or at least microbial, origin. Fuller (1947) obtained some suggestive indications that the uronic compounds of soil were of bacterial, not plant, origin.

It is not, however, yet certain that these still vaguely defined uronic compounds are the actual gums that enable humus to do its main job in forming and preserving a soil structure. Much attention is now being paid to them, so far with relatively little result. They are universal and very characteristic constituents of soil humus, and may comprise up to a quarter of the total soil carbon (Shorey and Martin, 1930). The proportion of uronides seems to increase with the depth at which the humus is sampled, but the significance of this is not known. Attempts have been made, with some success in the laboratory, to create a crumb structure simply by adding uronide substances to soil. Alginic acid, a uronide obtainable from seaweed, and some of its salts can be made to produce quite a good crumb structure, apparently simply by glueing the soil particles together, but it does not work very well on a field scale.

At the end of 1951 the Monsanto Chemical Company announced that it had developed a synthetic resin—'krilium'—which is said to be resistant to microbial attack and to produce a very stable and enduring crumb structure in soil. Krilium is not a uronide, but it seems to act in a similar way in glueing soil particles into aggregates. Whether or not its use will make any big differences to ordinary agricultural practices remains to be seen, but, being a definite substance of presumably known constitution and obtainable in large quantities, it should be able to provide us with much information about the actual mechanism of crumb formation. The very fact that a substance has been made in a chemical factory that will perform the structure-forming function of humus, is a sign that man is progressing towards being able to take over all its functions and so reduce his dependence on microbes and worms and their excrement.

What Humus Does

Most of the studies which have been made, and in my opinion largely wasted, on humus have been concerned with what humus *is*—is it essentially a ligno-protein complex, how are the nitrogen, phosphorus and iron combined, what is the nature of the polysaccharides, and so on ? These things will have to be explained some day, but meanwhile it seems more urgent to find out exactly what humus *does*. We know, for example, that a large part of soil humus consists of lignin-like compounds, in all probability derived from the indigestible and microbially resistant lignin of plants, but we do not know whether these lignin derivatives represent the business part of humus (as they well may) or the biologically inactive skeleton to which active groups and compounds are attached. We suspect strongly that the polysaccharide gums are active in aggregate formation, if so, it seems more important to discover how they act than what they are. Humus has for long been regarded with something approaching reverence—well deserved because since the beginnings of agriculture farmers have found that many of the worst failings of their soils could be remedied by keeping them supplied with humus. If we could discover the mechanism by which humus confers its many blessings the reverence would disappear, which would be a pity because other intangible values would go with it, but we should have gone a long way towards solving one of the biggest practical problems confronting soil science : how to produce more from an acre with less humus.

Waksman (1936), in his monumental work on humus, summarises what humus does—it improves the soil physically and modifies its biological nature, producing conditions favourable to plant growth. The statement requires qualifying because humus in peat soils makes conditions favourable for the growth of peat-loving, but not agricultural, plants, and in most uncultivated soils it favours the growth of the plants already there against potential invaders. This is one of the reasons for the stability of plant associations ; the acid humus of a podzol produces a weakly developed, laminated soil structure, very unsuitable for agricultural and most other plants, but suitable for conifers and their associates, which can thus hold their own against more exacting, potential competitors. Here, however, we shall be concerned chiefly with the humus of agricultural soils. Let us analyse in more detail how it improves the soil and modifies its biological nature.

Humus and humification—two quite distinct concepts—do the following things of significance in agriculture.

1. Humification provides a supply of nitrogen, phosphorus and several other plant nutrients in available form. The nutrients are stored in organic combination, and are mostly quite unavailable to plants until they are released during decomposition of the organic material. Since decomposition is promoted by much the same forces as promote plant growth this means that plant nutrients are protected from being washed out of the soil when they are not wanted, and are presented to the plants ready ' cooked ' when they are wanted. The whole process is biologically very efficient—a necessary characteristic under ' natural ' conditions where there is never anything to spare.

2. Humus itself does not provide organic food for anything except, perhaps, a few rather exceptional micro-organisms with depraved tastes. It is the end-product of the bodies of innumerable different organisms, great and small, being eaten, digested and excreted by others. In the process of humification, however, much biological energy is released and many useful by-products (such as mineral plant nutrients and carbon dioxide) are formed.

3. Humus can combine, under certain conditions, with clay minerals forming a partly organic, partly inorganic complex (the clay-humus complex) of which the soil crumbs, which are indispensable for agricultural soil fertility, are composed. The humic part (or part of the humic part) acts as a glue binding the mineral particles together, but it is by no means the only factor in crumb formation.

4. In the form of the clay-humus complex the microbially very resistant humus seems to undergo a kind of ' autoxidation,' in which it must disappear completely since the more resistant (lignin) parts do not increase at the expense of the

less resistant (protein and carbohydrate) parts. In the process plant nutrients must be released, but it is not known what proportions of the total organically combined nutrients are released during humus formation and humus oxidation, respectively.

Thus plant and animal residues, in their passage from the fresh state, through soil humus, to thin air and a pinch of salts not only provide a complete and self-regulating supply of nutrients for any plants growing on the soil, but also promote the essential physical conditions in which plant roots can thrive. These conditions vary according to the nature of the plant, and the humus derived from one plant community and its associated fauna tends to produce physical soil conditions favouring the survival of that community. The conditions required by all agricultural crops include good aeration, water-holding capacity, permeability and drainage—more exacting conditions than those required by any wild-plant community. Decomposing plant and animal residues, but not humus, also provide a very favourable medium for the development of micro-organisms and lesser animals, but the farmer is concerned not to foster these organisms, but to obtain and use the pinch of salts left when they have done their work.

Humus on the Farm

Perhaps the most significant feature about the functions of humus in soil economy is that they are regulated by the same forces as influence plant growth, and thus come into operation as and when they are needed by the community living on the soil. Both the job of supplying plant nutrients and the job of maintaining favourable physical conditions in the soil can be performed by man—in some respects better, in some respects worse, than by humus—but man's actions are not regulated so nicely by seasons as is decomposition in soil. Consequently crops are liable to receive man-given inorganic fertilisers, which are available immediately on application, at the wrong times and in the wrong quantities. This is one of the main reasons for the continuing popularity of organic manures, in spite of their inconvenience and high cost ; they are practically fool-proof. The farmer puts them on and Nature does the rest. But when using inorganic fertilisers he has to do a lot of thinking, and even then, as likely as not, unpredictable weather will upset his calculations. Science is, however, progressing towards more rational and controlled application of fertilisers, by means of machines for placement, improvements in physical condition, and invention of new fertilisers, and each step forward means that the plant-feeding operations of man are approaching nearer in biological efficiency to those of humus, which they have already exceeded in economic efficiency.

As far as the provision of plant nutrients is concerned, man has made himself practically independent of the process of humification in every industrialised country in the world. Although the world is still far from having enough of all the fertilisers it needs for adequate food production, there is no shortage of the raw materials for their manufacture, and in most industrialised countries they are much cheaper, more plentiful and more convenient to use than organic manures. Artificial fertilisers, however, suffer from the serious drawback already mentioned that the supply to the plant of the nutrients they contain is not automatically regulated according to the needs of the plant. The automatic regulation of organic manures is by no means perfect, particularly for agricultural crops, but it is better than nothing, and gives the average user more confidence in organics than in artificials. At present there is no likelihood of discovering economic inorganic fertilisers that will be so *biologically* efficient as organic manure, but much will doubtless be done to improve the reliability and economic efficiency of artificials. The farmer can best improve on the so-called humus manures by using them together with artificials, thus getting the advantages of both. The fact that most organic manures are ' balanced ' manures does not mean that they are necessarily the best balanced for all crops.

It has hitherto been more difficult for the farmer to find substitutes for the physical effects of humus. Humus, being a colloid, increases the absorption capacity of soil—that is, its capacity for absorbing plant nutrients from solution and keeping them in the soil, and also its water-holding capacity. These qualities are particularly valuable in sandy soils, the absorption and water-holding capacities of which are low. In clayey soils, which are already highly colloidal, humus has a lightening effect, increasing permeability and porosity. To have the greatest effect lime is required, and the effect is probably produced by the formation of the clay-humus complex which has physical properties much more favourable to agriculture than either clay or humus separately. The clay-humus complex is the chief component of the water-stable crumbs without which continuous agriculture is impossible. Humus is, indeed, about the only substance known that will cement soil mineral particles into stable crumbs, although it may be found that synthetic substances like krilium (p. 108) can be equally or more effective.

Sekera (1951) believes that the soil crumbs consist of primary aggregates consisting of separate soil particles cemented together by soil colloids, and secondary aggregates or crumbs consisting of primary aggregates held together by products of the activity of living organisms. According to this view, humus, as distinct from humification, is concerned only in the creation of primary aggregates, which are mostly very small, only a fraction of a millimetre in diameter. They are too small to be affected either favourably or adversely by cultivation, and there is little a farmer can do at present to replace the glueing action of humus except to supply more humus. One of the most serious results of a lack of humus is that, in certain climates, soils become liable to water and wind erosion when the primary aggregates break down to powder. The tendency to erode can be countered by mechanical measures like the construction of terraces and windbreaks, but the loss of structure cannot. Particularly in semi-arid countries loss of humus following the ploughing up and cultivation of natural grasslands has resulted in widespread soil erosion. The best remedy is to put the soil back to grass until its humus content and structure are restored.

The formation of secondary soil aggregates which, according to Sekera, are made by the actions of living organisms—the compressing action of plant roots, the binding action of fungal mycelia and the digesting action of earthworms and other animals and bacteria—can to some extent be replaced by cultivation, which is discussed more fully in Chapter VIII. Some of the less violent measures like harrowing, rolling and hoeing are designed to break up clods or compress smaller aggregates into crumbs. Such mechanically made clods are not so water-stable as those made by biological means, but if the right measure is taken at the right time, a soil structure can often be made mechanically that will serve its purpose sufficiently long to enable a growing crop to obtain a sufficiency of nutrients, moisture and air from the soil.

The farmer is thus still a long way off being able to dispense with humus. The two functions of humus that interest the farmer most are the provision of plant nutrients and the maintenance of soil structure. The farmer in Britain can now obtain more plant nutrients more cheaply from artificials than from organic manures, and if his soil is in good heart the artificials will work the more efficiently. But to ensure good heart or good structure in the soil he will require to maintain the humus content. In most British soils the humus seems to be rather resistant to oxidation, and the necessary replacements to recoup annual losses of humus from cultivated soils are often provided by the roots and stubble of a grain crop. The growing English practice of following the Scottish practice of ley farming should provide ample humus to maintain soil structure, and there is no virtue in overfeeding soils with expensive and wasteful organic matter. Whether synthetic substances like krilium will be able to take the place of humus as a structure stabiliser remains to be seen ; if so we shall have gone a step further towards freeing agriculture from dependence on inedible organisms like earthworms, ants, fungi and bacteria, and shall be able to concentrate more on growing tastier things.

But the most important function of humus, or more correctly, of the whole organic system of the soil, is to act as a regulator of soil processes, especially of plant growth. The same external conditions of temperature and moisture that encourage plant growth also stimulate earthworms, bacteria and fungi into activity, which not only provides a rich supply of plant nutrients but also substances which will help to pre-preserve soil structure throughout the growing period. It should be remarked, however, that this efficient regulation of soil processes does not operate solely for the benefit of plants ; the activities of plants, earthworms and bacteria help, or check, the soil fungi—and so on. The agriculturist's ultimate goal should be to produce economic plants only, but until he can do so he will have to put up with humus. Agriculture does not become less natural as humus becomes less indispensable to soil fertility, for man's brains and hands which try to do the job of humus are just as natural as an earthworm's guts even if, at present, they are not so efficient. They have not had such long experience.

CHAPTER VIII

CULTIVATION

An uncultivated soil is obviously in a condition to provide the plants growing on it with all they require in the way of nutrients, water and air, otherwise the plants would not be there. The consequent harmony in which a natural plant community lives has led many people to believe and assert that to copy Nature's balanced economy, which provides a sufficiency for all without impairing soil fertility, should be the goal of the agriculturist. Competition in nature, however, ensures that nothing gets more than the bare minimum for existence, whereas agriculture, except of the subsistence type, is essentially an affair of producing huge surpluses over what the land naturally yields or can be consumed by the natural fauna.

For most of human history the main measure for increasing the agricultural productivity of the soil has been cultivation. Cultivation does several things : it pulverises and loosens the soil, making it more permeable and promoting aeration and drainage, or it compacts the soil when it is too loose ; it destroys weeds both by burying them (when the soil is inverted to some depth), and by directly killing them (by surface cultivation) ; it creates a fine seedbed in which the majority (instead of, as in nature, a small minority) of seeds sown will germinate ; it mixes manures and fertilisers into the body of the soil ; and it provides a mulch of loose soil on the surface that reduces surface evaporation of moisture.

Cultivation has always, and rightly, been deeply respected by farmers. It is the first operation required to get rid of the natural vegetation and so to give the unnatural vegetation of agriculture a chance to grow ; it is symbolic of the back-breaking labour with which men have continuously fought, and seldom so hard as in Britain, to keep the natural vegetation out of their fields. But until quite recently nobody had attempted to measure its effects on plant growth. The tradition had become almost universally established that the more, and the more thoroughly, one cultivated the better for the crops. So firmly was the tradition established that a theory of soil-water movement—the capillary theory—was propounded to explain the benefits of cultivation that were taken for granted. Water was said to move in all directions through the soil by capillarity—that is, by the force which causes liquid to rise in a narrow open tube, or to be absorbed by blotting-paper or a porous brick. Soil is honeycombed by cells of all dimensions, wide and narrow. The effects of cultivation measures were to regulate capillary movements of water either by breaking the soil capillaries and so reducing the movement of water to, and evaporation from, the surface, or by constricting them and so increasing the soil's power of lifting water from a depth. The first effect was achieved by harrowing or hoeing, the second by rolling.

Capillarity could thus satisfactorily explain many of the alleged benefits of cultivation as it affected moisture content, but doubts were cast on the explanation when it was found that on both theoretical and experimental grounds the capillary theory was untenable. Arguments, based on sound physical theory, suggested that cultivation should have very little influence in controlling soil moisture. Doubts were then cast on whether cultivation, beyond that needed to clear the ground and prepare a seedbed, had any virtue, and it was found that the question had never been tested experimentally. Within the last twenty-five years, however, numerous carefully controlled experiments have been made in England by Keen (1942) and others that have shown that in a wide variety of soils, cultivations, over and above those necessary to prepare a seedbed and control early weeds, have little effect on crop yields. Similar measurements made in the drier conditions of the American prairies have given similar results. With occasional exceptions, operations like subsoiling, deep ploughing and double ploughing (autumn and spring) had little or no effect on crop yields, nor was there any appreciable difference between ploughing, grubbing and rotary cultivation. What did appear from these experiments was the value of cultivations for weed control. Indeed, a peculiarity of the plough mould-board, which was invented in Germany in the earliest days of north-western European agriculture, is its capacity for killing weeds in the damp cool weather of north-western Europe. No other implement has a comparable capacity. An added advantage of the mould-board at the time of its invention may have been that by inverting the plough layer every year it retained in that layer plant nutrients which might otherwise have been washed right out of the soil in the prevailing humid climate. Nearly all previous settled agriculture had been in drier regions, and had confined cultivation to scratching the soil surface.

The plough, the harrow and the hoe kill weeds which compete with the crop for moisture and plant nutrients in the soil. In some of Keen's experiments deep ploughing did appear to increase crop yields compared with those obtained after shallow ploughing, but in these experiments the land was dirty, and counts of weeds and weed seeds made during the experiments confirmed that deep ploughing did markedly depress the weed population. If the land is already clean, as it usually is after a root crop, very little cultivation beyond that required to produce a good seedbed tilth is required ; further cultivation is a waste of money and labour. Nearly always the most fertile soil is on the surface, and bringing up the subsoil by deep ploughing is bound to dilute the fertility. There may, however, be compensating advantages such as improved drainage or the breaking of a hardpan.

One reason why frequent and thorough cultivation has won what now seems to be an unwarranted reputation as a criterion of good farming is that it is mostly practised by good or, what frequently comes to the same thing, wealthy farmers. The high yields associated with thorough cultivation are usually due to a combination of favourable circumstances among which cultivation plays a minor part.

The benefits derived from rolling or otherwise compacting the soil were attributed in the days of the capillary theory to an intensification of capillarity enabling water to rise more easily through the soil from the water table, or to move more easily from a moist area to an area near a root, dried by root absorption. But under the most favourable conditions water will not rise by capillarity more than about two feet, which is less than the depth of most water tables, and water does not move appreciably through a soil at or below field moisture capacity (p. 55). Keen gave a simpler explanation of the effect of rolling. Rolling pressed the soil closer round the roots of young plants so that the growing roots found it easier to go after the water. (According to the older explanation the water went after the roots.) For any water movement through the soil of the kind implied in the capillary theory the soil would have to be so damp that rolling would be out of the question.

According to the capillary theory, and also to a very widespread belief, one of the chief virtues of hoeing, apart from the destruction of weeds, was the creation of a mulch of loose powdered soil on the soil surface. The network of soil pores was believed to act as a network of capillary tubes through which water rose from below to the surface where it evaporated; hoeing broke the capillary system and so retarded surface evaporation. According to present-day theory, however, which postulates that water in soil is mostly immobile, continuous upward movement from below to replace water lost by evaporation does not occur. An unhoed soil will dry out *completely*, layer by layer, from the surface downwards, and the water content of the moist layer below the dried surface will not be affected by the water content at the surface. Many soils, when they dry, form a loose powder and are thus self-mulching. In such cases hoeing, regarded as a mere cultivation and not as a weed-killing measure, is redundant, but in other cases where the soil dries to a crust or forms cracks

(1,030)   9 through which water can evaporate a hoe cultivation can have agricultural or horticultural value. Hoeing may do actual damage by destroying surface roots, though this may stimulate plants to develop deeper root systems and so to be able to tap deeper sources of plant nutrients and water.

All experiments made in Britain and America have shown the supreme importance of the weed-killing function of cultivation. It is clear that crops are liable to suffer heavily from weed competition, especially in the early stages of their growth, and once a crop has suffered such a check it will not usually recover from it. Cultivation is also useful in preparing a suitable seedbed, but it appears that the tilth of a seedbed is not so important as its freedom from weeds or its moisture content. Cultivation helps, too, to distribute fertilisers through the soil. Plant nutrients, since they move in solution, are as immobile in the soil as water, and stay where they are put. One advantage of deep ploughing that can sometimes confer great benefit is that it enables fertilisers to be incorporated at depth where plant roots can use them in periods of drought when the surface soil dries out.

The effect of the usual measures of cultivation on soil structure and its stability is generally deleterious, though here again few actual measurements have been made. As most people in Europe assumed that thorough cultivation promotes high yields so most people elsewhere have assumed—with rather more reason—that continued cultivation harms the soil. The strongest evidence supporting this assumption is the destructive erosion that the once beautifully structured black earths (chernozems) of continental grasslands have suffered after a few decades of cultivation. Most of the original soil crumbs have been pulverised so that the soils wash or blow away easily, but although cultivation is certainly partly to blame for this state of affairs, it has not been the only, or even the major, factor in destroying soil structure. The main factors have been the loss of soil fertility removed in crops and not replaced in manures, and the absence of a perennial-grass cover. The plough administered the *coup de grâce* to crumbs that were already feeling groggy. Soil crumbs in overgrazed pastures that had never felt the plough went the same way.

Probably the main effect of cultivation on soil structure is not so much a direct mechanical pulverisation of crumbs (though that does take place) as an enhanced oxidation of humus and consequent reduction in the amount of glueing material holding the soil particles together. Cultivation promotes humus oxidation by increasing soil aeration—that is, by its loosening a compacted soil. Operations like rolling would not reduce the humus content. The loss of organic matter when land is first broken is rapid, but then slows down, and in time the humus content approaches an equilibrium value. Measurements made in Australia showed that the initially rapid and then slower decline in the humus (or nitrogen) content of cultivated soil was paralleled by a similar rapid and then slow decline in the degree of aggregation (Clarke and Marshall, 1947). This suggests the existence of an active, easily decomposable humus, and of an inactive humus that only decomposes slowly. Actively decomposing organic matter is required for the formation and maintenance of soil crumbs, and, at least in temperate countries, the harmful consequences of cultivation on soil structure can be averted, as is commonly done in Britain, by applying organic manure occasionally or by putting arable land into grass.

Ploughing can even improve soil structure for agricultural purposes in heavy soils and make larger aggregates out of small aggregates in light soils. Whether or not these desirable effects are produced depends on the moistness of the soil at the time of cultivation. If a clay soil or a heavy loam is cultivated dry, it breaks up not into crumbs but into large hard clods, and in such a condition cultivation requires a wasteful expenditure of energy. If such a soil is cultivated when very moist, it breaks into compacted ' lenses ' with polished surfaces that again yield hard clods when dried. Overmoist cultivation requires even more energy than dry cultivation, and may be impossible.   The range of moistness within which heavy soils can be cultivated with benefit to their structure is narrow, the range widens as soil texture gets coarser, but depends on other properties, including humus content and type of clay mineral, besides texture.   The moisture range within which an aggregate structure can be produced by cultivation is

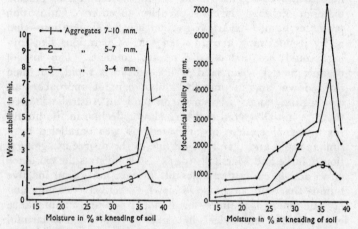

Fig. 3   Stability of aggregates developed by kneading soil at different moisture contents.   The peak indicates the 'moisture of structure formation'.

usually within the plasticity limits—that is, between the lowest and highest moisture contents at which a soil, when kneaded, will retain its shape.

Vilensky (1945) showed that for every soil there was a certain moisture content, which he called the 'moisture of structure formation,' at which the strength and stability of the aggregates formed by kneading the soil (in the laboratory) was a maximum from which they fell away quite sharply when the moisture was increased or decreased (Fig. 3).

The moisture of structure formation was generally a little below the upper plastic limit.   It varied from soil to soil, but averaged 30–40 per cent. Vilensky suggested that at the moisture of structure formation a soil was in the most suitable condition to benefit by cultivation. Cultivation could, however, only create structure if other factors—humus content, soil reaction, biological activity—were favourable. Every arable farmer knows the importance of cultivating soils, particularly heavy soils, when they are neither too wet nor too dry. If he catches his soil at the moisture of structure formation (which no laboratory will yet determine for him) he will derive most benefit from his work.

Even in a country with a climate as moist as that of Britain there is a probability that at some time during the growing season crops will suffer from lack of water. The higher the yields, the greater the demand on soil water. At times there is too much water, at times there is too little, and one of the purposes of soil management, and of cultivation in particular, is to store in the soil the greatest possible amount of water in the surplus season for use in the deficiency season—and with high-yielding crops there is nearly always some deficiency during the summer. A purpose of autumn cultivation is to loosen soil, which may have been compacted by trampling and the passage of implements during the preceding harvest, so that winter rains will penetrate it easily. Furthermore, a heavy soil containing much clay may be too moist after the winter to cultivate at the appropriate time in the spring, whereas autumn cultivation will open up the soil not only to water penetration, but also to the action of frost which in certain circumstances can be of great value in breaking down winter clods to a favourable crumb structure. A spring cultivation should then finish the process if carried out when the soil is at a suitable degree of wetness, though its main purpose is to bury weed seedlings.

Ploughing also exposes the furrow slice to the powerful weathering influence of frost. The effect of frost on soil structure is rather complicated, and it should not be assumed that, because autumn ploughing is often recommended so as to allow winter frosts to act on the soil, the results of such action are necessarily beneficial. The structure of sandy soils is liable to be destroyed by exposure to frost. The breakdown of a heavy soil to a stable crumby tilth can be greatly assisted by freezing, particularly if the freezing is gradual, as it usually is in England, and the moisture content is moderate (Russell, 1950). In these circumstances parts of the soil are left ice-free, and are subjected to compression by expansion of the ice in the frozen parts, whereby very stable compressed aggregates are produced. A high water content and repeated freezing at low temperatures produce a structure too finely granulated to be the best for agricultural purposes.

Slater and Hopp (1949) suggest that the apparent benefits to soil tilth resulting from the breakdown by frost of autumn-ploughed land are largely illusory. The frost action fractures the lumps and clods, but also destroys the bonds holding water-stable aggregates together. As a result the soil, while apparently in good condition because of its freedom from clods, has so little water stability remaining that the favourable tilth produced by cultivation and frost action is not retained throughout the season. It is difficult to prove or disprove such a statement because there may be other factors, like biological activity, which will restore a shattered structure during the growing season. In former days when labour was cheap the maintenance of at least a semblance of soil structure by frequent cultivations to keep the soil loose and open while a crop was growing was a recognised practice, and was the basis of Jethro Tull's (1674–1741) system of drill husbandry (p. 202).

Such cultivations are, however, very laborious, and today would be a very expensive method of doing what a stable soil structure will achieve far more efficiently. Tull's experience merely shows that human labour and ingenuity can always find a substitute for the operations of Nature. In a similar way, recent and future developments in herbicides may make the weed-killing function of cultivation superfluous, but they

have not done so yet.   Most of the good results Tull got by
his cultivations are now attributed to the killing of weeds.

The triumphant success of the mould-board plough in holding in check the most aggressive pests of Western European
agriculture (even though its success was not recognised in that
light) has led to its adoption in every other continent colonised
by Europeans.   In some circumstances, however, weeds can
be as much friends as foes of the farmer, particularly where
erosion is liable to occur and it is necessary to maintain a
protective cover of vegetation for as long as possible.   The
mould-board plough, which leaves the upturned soil bare of
vegetation, can then be a dangerous implement.   But such
a well-established favourite is not easily abandoned, though
many attempts are being made to find alternative methods
of cultivation that will kill weeds without depriving the soil
of the protection which plants afford against erosion.   The
tendency is to revert to or modify the pre-European methods
of cultivation that merely scratched the soil surface.   A successful method which has been widely adopted in North America
is that known as the ' ploughless fallow ' in Canada and as
' stubble mulching ' in the United States.   Weeds are under-
cut by using a disc plough or, more effectively, a special
cultivator which cuts the soil in a plane parallel to, and a few
inches below, the surface.   Weeds, and stubble and residues
of the preceding crop, are left anchored on the surface, and
the soil is disturbed as little as possible.   The residues break
the force of rain and otherwise protect the soil until they die
and decay, when they make humus which ultimately gets
incorporated into the surface soil and gives it some additional
resistance to erosion and greater capacity for absorbing and
conserving moisture.   The method was originated to protect
the soil during the fallow year, which is still unavoidable in
the dry Canadian prairies, to store up moisture, but it is equally
effective where cropping is done every year.   Very high claims
have been made in the United States (Hendrickson, *et. al.*,
1943) for its value both in conserving soil and in maintaining

crop yields, and it has been described as 'a revolutionary farm practice.' A comparison between mould-board cultivation and stubble-mulch farming, with their diametrically opposed purposes, illustrates well that there are no universal principles of soil cultivation, but that the principles must be adapted to each combination of soil and climatic conditions.

Several people have advocated the elimination of all cultivation except for the preparation of a seedbed (Faulkner, 1943 ; Smith, 1950). The idea is the same as that of stubble mulching, namely, to keep all plant residues on the surface, to allow them to decay there and finally to be incorporated as humus in the uppermost layer of soil. Very impressive claims have been made by these non-cultivators for the good results of their methods, and there is little doubt that many of the claims are justified. They are, however, methods which at present are more suited to a garden or smallholding than to a commercial farm. They rely essentially on earthworms to do the necessary cultivation and to mix the organic residues with the soil, and earthworms, if they are present in adequate numbers, can perform these functions sometimes as well as, and sometimes better but always more slowly than, a ploughman. For large-scale operations and those which have to conform strictly to a time-table the ploughman is usually a more reliable agent than the earthworm, even though his final result may be less satisfying to the aesthete. Nevertheless, recent experience with stubble mulching and similar systems has demonstrated that in certain circumstances, particularly in arid and semi-arid regions, agriculture can be profitably organised so as to reduce cultivation to a low minimum. Always its function is to kill weeds.

We are now learning quite a bit about the physics of cultivation, and can discuss with understanding how cultivation affects and is affected by such soil properties as capillarity, pF, field moisture capacity, plasticity limits, hygroscopic coefficient and others, but this knowledge has been only recently acquired, and there has not been time for it to be used in modifying ordinary farming practice. It is questionable, indeed, whether it will ever be much used. Our methods of cultivation have become firmly established as a result of a thousand years of working the soil, and they are as much a part of our British way of life as a taproot is part of a pine tree's existence. The most our new knowledge is likely to do is slightly to accelerate the process of agricultural evolution, or perhaps merely to explain it. Whether or not science proves that three-quarters of our cultivations are superfluous we shall probably continue to make them fundamentally in the same way as we always have, occasionally using new implements and substituting a petrol engine for a horse, or plutonium for petrol. In younger countries there are greater possibilities and greater scope for change.

Many of the basic facts of cultivation, without the rather repellent but useful scientific terminology now attached to them, were well known four hundred and more years ago. Sir Anthony Fitzherbert, whose ' Boke of Husbandry ' (1523) is the earliest printed English book on agriculture, wrote about cultivation :

' Nowe these housbandes have sowen theyr pees, beanes, barly, and otes, and harowed them, it is the beste tyme to falowe in the latter ende of Marche and Apryll, for whete, rye, and barley, and lette the husbande doo the beste he can, to plowe a brode forowe and a depe, soo that he turne it cleane, and lay it flat, that it rere not on the edge ; the whiche shall destroy all the thistles and wedes : for the deper and the broder that he gothe, the more new molde, and the greatter clottes shall he have, and the greatter clottes the better wheate, for the clottes kepe the wheate warme all wynter, and at Marche they wyll melte and breake, and fal in manye small peces, the whiche is a newe dongynge and refresshynge of the corne ; and also there shall but lyttle wedes growe upon the falowes that are so falowed : for the plough goth undernethe the rootes of all maner of wedes,

and tourneth the roote upwarde, that it maye not growe.
And yf the lande be falowed in wynter tyme, it is farre the
worse, for thre principal causes, one is, all the rayne that
commeth shal washe the lande, and dryue away the donge,
and the good moulde, that the lande shall be moche the
worse.   Another cause is, the rayne shall beate the land so
flat, and bake it so hard togyther, that if a drye May come,
it will be too harde to stere in the moneth of June ; and the
thirde cause is, the wedes shall take suche roote er sterynge
tyme come, that they wylle not be cleane tourned undernethe, the which shall be great hurte to the corne, whan it
shall be sowen, and specially in the weding tyme of the same ;
and for any other thynge, make depe holowe forowe in the
rydge of the lande, and loke well, thou rest balke it nat,
for if thou do, there wyll be many thistles ; and than thou
shalte not make a cleane rydge at the fyrste sterynge, and
therefore it muste nedes be depe plowed, or elles thou shalt
nat tourne the wiedes cleane.'

CHAPTER IX

FOREST SOILS

THE agriculturist's prime aim in soil management is to produce and preserve a crumb structure. We cannot define precisely what we mean by the ideal crumb structure, and probably shall never be able to do so because soil structure is an indefinite concept, but an experienced agriculturist knows when his soil is in good heart, and the soil physicist knows that such a condition is invariably associated with certain physical properties of the soil. In Britain soil structure, whether crumb or otherwise, is mainly the result of the action of living things, among which man is by far the most active on agricultural soils. Crops and other plants, animals and micro-organisms are perhaps more direct agents in structure formation, but their activities are mostly subordinate to man. He is the dominant species who determines the ultimate properties of the soil, as trees are the dominant species in a forest and likewise determine the structure of the soil.

Man is seldom so completely dominant on a forest soil as he is on an agricultural soil, because trees are permanent crops that may occupy the soil for century-long generations. Very often they are the natural climax vegetation, and man's interest is to secure and strengthen their dominance over other living things. If he is afforesting an area for the first time he may have to see that the soil conditions are modified by cultivation or manuring to suit tree growth, but usually, when once a tree canopy has formed and the young trees have assumed dominance, they will themselves look after the soil to their own advantage and make it a suitable habitat for other organisms that will help in producing a forest soil. At present the Forestry Commission is planting with pines and spruces large areas of very acid moorland on which the natural vegetation is heather. A heath soil is a kind of podzol not unlike a coniferous-forest soil, nevertheless, young conifers can scarcely survive on it unless the heather is removed and kept from returning, because the heather has made a soil better adapted to itself than to conifers, which cannot compete with it. The best-known way of destroying the heather is to bury it by deep ploughing, but this will not prevent it from reoccuping its own ground in a few years' time. Consequently the young trees are encouraged to make rapid growth by fertilisers and other silvicultural measures, and if these are successful and the trees make a good start they will be able to hold their own until their branches form a closed canopy over the soil. Thereafter they become the dominant species, the heather is suppressed and the soil is gradually converted from a ploughed-up, mutilated heath podzol to a forest podzol.

Or that is what we hope and expect to happen. Some time during the next century our grandchildren should be able to verify whether it will have happened. The forest podzol is not unlike the heath podzol, and has already been described on p. 10. It differs from the heath podzol in that the latter often has a very characteristic thin hardpan, about an eighth of an inch thick, at the base of the bleached layer (Plate 3b). This pan is found in many British heath soils, and seems to be associated with the soil-forming activities of heather. How or why it is formed is not known, but its formation must be connected with the physico-chemical properties of the compounds of iron and humic acid produced when humic acid from the heather peat attacks the iron compounds in the soil. These are washed out of the topsoil by percolating rainwater, and are precipitated lower down as the hardpan. Similar, but not identical, iron compounds are produced by the action of coniferous-forest humus, but they are precipitated in a much thicker and less hard layer.

The pan of a heath podzol offers a serious impediment to tree growth, as it is almost impenetrable by tree roots and interferes with the free drainage of the soil. Besides destroying the heather, one of the main purposes of deep ploughing is to break up the pan. As it has not been found under forest growth it is hoped and expected that it will not reappear as the forest podzol develops. The absence of the pan will be the most obvious change that the forest will have effected in the soil profile, but from the trees' point of view it will be a decisive change. It makes all the difference between a habitat pre-eminently suited for heather roots and one better suited for the roots of coniferous trees. Ovington (1951) has described the initial changes which have taken place on an afforested sandy heath soil carrying 19-year-old pine trees. There has been a fall in the level of the water table, caused by the increased transpiration of the trees, producing a larger effective soil depth and better aeration. The original heather raw humus has been replaced by less acid pine humus with a greater nutrient content. The nutrient content of the sandy soil itself has, however, decreased, possibly owing to the greater demands made on nutrients by pines in comparison with heather.

Soil structure in the topsoil of a forest podzol is almost non-existent. There is practically no humus in the bleached ' A ' horizon to cement the mineral particles together, and there are very few plant roots to compress them together. Sometimes a faint horizontal lamination can be seen in the ' A ' horizon, but it is not known what causes this. The underlying brown or black ' B ' horizon is hard and compact, the mineral particles being cemented together by the humic acid and other substances washed out of the ' A ' horizon and precipitated in the ' B ' horizon. Most of the plant nutrients —especially phosphorus, calcium and potassium—are also removed from the ' A ' horizon, and much of them passes through the ' B ' horizon and is lost in the drainage water. Thus a podzol has neither the physical nor the chemical properties of an agricultural soil and is, indeed, almost useless for agriculture without drastic improvement. It is, however, the soil that pines, spruces and heather like best, not because of its inherent infertility but because they can tolerate the conditions they produce whereas more exacting types of vegetation that might compete for the soil cannot. Pines and spruces cannot indefinitely hold their own on soils rich in plant nutrients, of which they require comparatively little, and with a good structure. These virtues help their potential competitors more than themselves.

If, however, man is prepared to take the trouble and expense of protecting the conifers from competition he can get much better growth on a fertile than on an infertile soil. A virgin forest is a community in which the survival of the community takes precedence over that of the individual ; it does not matter how misshapen and starved the trees are so long as the unity of the forest is preserved. In a silvicultural forest, on the other hand, the quality of the trees is of great economic importance, and the forester directs his attention towards improving that quality even if, by so doing, he has to disturb the stable plant-soil equilibrium attained by a virgin forest. For example, he may—and often does—aim at retarding or even reversing the natural process of podzol formation under coniferous forest, and he can thereby effect a marked improvement in the stand, that his son or grandson may live to see. His aim in soil management is a long-term one, and he cannot, except with very young trees, expect his crop to respond dramatically or immediately to soil treatment.

Measures like cultivation and manuring are usually out of the question in a fully grown forest. The forester recognises that the trees are still the dominant influence on soil formation in the forest he controls, and he aims to induce the trees to produce a soil that will give not the greatest ecological security to the forest, but the greatest yield of commercially valuable timber consistent with preserving, with man's assistance, the survival of the forest. The chief instrument through which the forest acts on the soil is the humus formed from dead branches, leaves and needles by the soil fauna and microflora. It is by inducing changes in the humus that the forester can effect gradual changes in soil properties that will ultimately affect the quantity and quality of his yield.

In temperate forests there are two main types of humus named *mull* and *mor* by the Danish scientist Müller (1887), who first distinguished them. They have already been mentioned on p. 70. Mull occurs mostly under deciduous forest and is largely composed of the excrement of animals and insects ; mor occurs under coniferous forests, and is the product of the action of micro-organisms, especially fungi, on forest litter. In mull, humus and mineral soil are inextricably mixed in the passage of plant residues and soil through the bodies of soil animals, particularly earthworms. The surface of a mull soil consists largely of organic humus mixed with a little mineral matter, but the proportion of humus falls rapidly with the depth until at a depth of, say, 18 inches the soil is almost entirely mineral. There is no sharp division between the mainly organic and the mainly mineral soil. The whole soil has a quite well-developed crumb structure and gives an immediate though occasionally deceptive impression of richness.

In mor soils (a podzol is a typical example) the humus layer (mor) lies on the top of the mineral soil and is quite distinct from it. It forms a peaty layer usually known in this country as raw humus. Very few animals inhabit mor (which is strongly acid), this being the reason why the humus does not get mixed with the mineral soil. The microbial population of mor is also small, and consists predominantly of acid-tolerant fungi which effect a slow humification of the forest litter. Mor gives an impression of extreme poverty ; it is, indeed, poor in mineral plant nutrients which are released only slowly by decomposition of the humifying material. The slow rate of turnover of nutrient material is a main factor limiting tree growth on soils of pronounced mor character.

The forester can induce a more rapid tree growth in mor soils by accelerating the decomposition of the raw humus, which is the source of all the nitrogen and most of the phosphorus and potassium required in the forest economy. Němec (1929), in Czechoslovakia, has shown that the practice of removing forest litter (for cattle bedding) has a bad effect on tree growth. Accelerating the decomposition of the litter, on the other hand, has a beneficial effect. It can be brought about by making conditions in the humus layer more suitable for the development of bacteria and humifying animals. Reducing the acidity by liming is one way of doing this, but very large quantities of lime would be required, often over very large areas, and it is not at present regarded as economic. Letting in light to the forest floor by judicious thinning accelerates humus decomposition, but perhaps the most effective way, where economic and silvicultural considerations allow it, is to change the nature of the humus-forming material by introducing deciduous trees among the conifers. The leaves of many deciduous trees are richer in lime than are the needles of conifers, and produce a less acid humus in which earthworms and bacteria can live. Lime is extracted from the subsoil by the deeper-lying tree roots and ultimately deposited on the soil surface in the leaf fall, thus enriching the surface soil in lime. In a mixed deciduous-coniferous forest the raw humus is ' milder,' and assumes a character midway between mor and mull. The conifers tend to grow better than in pure stands, but of course there are fewer conifers to the acre, and the exploitation of a mixed forest is more difficult than of a pure forest.

A remarkable effect of the supersession of coniferous forest by birch is sometimes observed. The birch may come in spontaneously when pine or spruce forests are cleared by felling or by fire, and may form a complete cover under which conifers will regenerate naturally and finally over-top and suppress the birch which will disappear more or less completely. But while the birch is dominant it can transform the soil from a podzol to a brown forest soil. This transformation involves the disappearance of the bleached podzol horizon and the underlying compacted horizon and the merging of the

two into a mull layer. It is brought about by the birch roots
extracting from the subsoil lime and other substances which
find their way via the mild birch humus into the topsoil.
Animals and bacteria, invading the humus layer, transform
it into mull. The transformation is seldom complete before
the resurgent conifers oust the birch and with their more acid
humus start podzol formation again. The cycle conifer-birch-
conifer represents a natural regeneration of the soil, since the
brown forest soil has a better structure and is richer in available plant nutrients than the podzol. But conifers, though
they thrive better in a brown forest soil, change it back to a
podzol because they are safer there from competition with
deciduous trees.

Attempts have been made in Scandinavia and elsewhere
to control the cycle so as to derive the greatest benefit to the
soil from the incursion of birches, but the difficulty in this and
many other forest-soil problems is that in a forest cycle lasting
for at least one and probably several centuries it is impossible
to state what is the greatest benefit to the soil. A mull humus,
however, gives more aesthetic satisfaction than a mor humus
to the forester as well as to the agriculturist, and there is no
reason why their instincts should be less reliable now that
mulls and mors have been named, classified and dissected
than they were beforehand.

The chemical properties of forest soils are not so significant
for timber production as those of agricultural soils are for crop
production—in spite of the fact that chemical deficiencies in
forest soils usually cannot be remedied, whereas in agricultural
soils they can be. Forest trees make very small demands on
the nutrients in the soil compared with those made by agricultural crops, and as they have much more extensive root
systems they can satisfy their smaller requirements from a
larger volume of soil. Nutrient deficiencies seldom arise in
a forest on any kind of soil, except when trees are planted
out very young and have not had time to develop an adequate
root system. Von Falkenstein (1911) found excellent (Class I)

(1,030)   10 and poor (Class IV) pine stands on soils of identical chemical composition in Germany.

Nearly all of the nutrients taken by trees from the soil go into the leaves, needles and bark, and are soon returned to the soil. Timber may only be removed once in a century, and then only small quantities of nutrients go with it. The nutrient economy of a forest which is being rationally exploited can therefore look after itself provided that the litter is left on the forest floor to humify.

The following are the amounts of nutrients, in pounds per acre,[1] contained in the annual leaf fall of beech, spruce and pine forests (Němec, 1929).

	Beech	Spruce	Pine
N	33	50	40
P_2O_5	10·5	6·4	3·7
K_2O	9·9	4·9	4·8
CaO	81·9	60·9	18·9

These figures should be regarded as very approximate, and they would vary greatly with the age of the stand, nature of the soil, etc. The smaller quantities of mineral and basic nutrients (P_2O_5, K_2O and CaO) in coniferous litter should be noted. Coniferous litter contains more nitrogen than beech litter, but it is less readily decomposed and the nitrogen becomes available more slowly.

Most of the working nutrient capital of a forest is thus locked up in the litter and humus, and for the healthy survival of the forest it is important that the litter should not be removed. Falls of 50–80 per cent in amounts of available nutrients in the soil, as determined by chemical analysis, are found after a few years of litter removal, and corresponding falls occur in the nutrient content of leaves and needles (Němec, 1929). Thus the litter and humus themselves become impoverished, so that even if litter removal is stopped it takes some time for the forest to recover its nutrient balance.

The soil characteristic which most commonly has special

---

[1] In kilogrammes per hectare in the original

significance for forest growth is permeability to water. Soils carrying forest cannot be artificially underdrained. For this reason, many trees seem to thrive best on light or sandy soils. Certain exacting trees, like beeches, do well on heavy clay soils which have higher nutrient contents than sands, but the humus from these trees produces an open soil structure which improves porosity and permeability. Conifers, on the other hand, can satisfy their small mineral requirements from sandy soils which are permeable even if, as often happens under conifers, they are structureless podzols. A heavier subsoil to prevent excessive drainage and hold water within the root range seems to be advantageous (Coile, 1935). By creating the compact indurated horizon of a podzol conifers themselves make a heavy subsoil.

In most countries practical problems of forest soils are concerned primarily with maintaining or increasing the yield, expressed as the annual increment of timber growth, over a long period of time. This is by no means solely a question of keeping the soil in good condition, but in so far as it concerns the soil the first consideration is to ensure that the soil remains a *forest* soil. Where clear-cutting is involved there is, of course, no question of maintaining the yield, which is obtained all at once and finally when the forest is cut down. Modern forestry practice, however, favours on both economic and silvicultural grounds some system of selection felling in which the forest comprises trees of all ages up to, say, sixty years old. At each felling only the oldest trees are cut, which, if the age classes are properly mixed, can be done without destroying the ecological unity of the forest. No large breaks are made in the tree canopy, and small breaks are soon filled by the remaining growing trees. Selection felling demands great skill in forest management, but it ensures that the biological influences—the tree canopy and the associated flora and fauna —that made a forest soil will continue to operate while the forest is being exploited. If the forest qualities of the soil do not deteriorate and natural regeneration of the desired tree species can be effected, fellings may be made continuously and the forest maintained indefinitely. Regeneration is clearly essential to the maintenance of the forest, and much of the art of selection felling consists in arranging that the right amounts of light, warmth and moisture reach the forest floor to allow tree seeds to germinate and the seedlings to establish themselves.

It is impossible to lay down any general rules for securing natural regeneration, as each species requires special conditions. So far as the soil is concerned, an actively decomposing humus layer, indicating a productive biological environment, is advantageous. The forester can and does modify the humus conditions by the system of felling he employs, but the system is governed more by accumulated empirical experience and the art of the forester than by scientific principles.

In Britain the proportion of land under forest is lower than in any other natural-forest country except perhaps Portugal. The typical British attitude to food production—that it is the foreigner's job—is even more strongly expressed with regard to timber production. Imports of softwoods amount to over 90 per cent of consumption, so the felling of the greater part of the country's marketable timber during the last war made little impression on the public, which still vociferously censures the Forestry Commission for planting useful rather than decorative trees on our wastelands. The main forest-soil problem in Britain is not to maintain fertility, but to get trees to grow and make a forest soil on land which is so poor that nobody wants it. Large-scale afforestation has only been done for some thirty years, and there is still much to learn.

We have already described how certain heath soils have been successfully afforested by deep-ploughing the soil to bury the heather and break the heath-soil hardpan. On these and other different soils the important thing is, firstly, to get the young trees, which are planted out when two to four years old, established and, secondly, to get a closed canopy formed as quickly as possible so that forest-soil conditions are created before the natural vegetation—of heather, rushes, sedges, upland grasses, etc.—gains the upper hand and suppresses the trees. The acid, upland peats, which comprise one of the most extensive and difficult land types that are being afforested, are very poor in plant nutrients, and young conifers, reared in the relative luxury of a forest nursery, tend to ' go into check ' and cease growth as soon as they are planted out. The needles turn yellow, suggesting nitrogen starvation, and the trees may remain for ten years or more without showing perceptible development and then begin slowly to grow. Many, however, succumb while in check.

A remarkable stimulant to growth on acid peats is given by dressing the tree seedlings, when they are planted out, with a little basic slag—a phosphate fertiliser containing lime which helps to neutralise the acidity of the peat. Slagged plants usually go right ahead, produce dark green needles and look well nourished while unslagged plants on the same ground remain completely in check. It is not known how the slag acts, but probably the combination of lime to neutralise acidity and phosphate to stimulate microbiological activity promotes decomposition of the peat and release of plant nutrients, particularly nitrogen. Nitrogen in undecomposing peat is quite unavailable to trees. The effect is cumulative ; plants which have received two ounces of slag at planting may develop into full-sized trees while unslagged plants grow only a few feet high. The early fillip given to slagged plants enables them to develop a good root system and get sufficient nourishment from a larger area of impoverished peat. It would have been impossible to afforest many of our peat areas, which are now growing promising young trees, without basic slag.

Although conifers can tolerate wetter and generally harder conditions than agricultural crops, drainage is often necessary on peat soils. Drainage is usually done by a system of open ditches, since tree roots are liable to interfere with tile drains.

A microbiological factor which seems to have more practical

significance in forest than in agricultural soils is the association
of tree root and symbiotic fungus known as mycorrhiza,
described on p. 94. Under difficult conditions mycorrhizas
can be of great benefit to young conifers, and inoculation of
the soil with the appropriate fungus may produce marked
increases in growth. The cause of the response to inoculation
may be similar to that to basic slag application—an extended
root surface enabling the trees to extract more nutrients from
the soil. Or it may be something quite different. In most
established forests mycorrhizas are almost universal on the
tree roots.

CHAPTER X

PLANT-MADE AND MAN-MADE SOILS

LET us now try to analyse how living communities act upon the soils they inhabit. It must be admitted, to start with, that very little is known about it, but if we study three strongly contrasted types of community—a coniferous forest, a grass steppe and a cultivated farm—we can form a general picture of the different ways in which each living community affects the properties of the soil. The soil, however, is not passive. In its turn it influences very strongly the nature of the fauna and flora. Soil, fauna and flora are each the result of the action of the other two.

The influence of the higher animals (except man) on soil formation is small owing to the fact that they move about and play an ephemeral rôle in the community's economy, but occasionally an animal species will become the dominant member of the community, like the rabbit in parts of Australia, or, nearer home, in the Suffolk Breckland. It may then transform the whole vegetation and thus indirectly affect soil formation. In considering the influence of a typical forest or steppe on soil formation, however, we can usually ignore the large fauna. For the same reason we can also ignore insects except those actually inhabiting the soil, which collectively have a considerable influence on soil-forming processes.

FOREST

A coniferous-forest society consists essentially of the dominant trees, a ground flora of herbs and mosses, a relatively small faunal population of animals, soil-inhabiting insects and slugs, and a micro-organic population consisting predominantly of fungi. The climate is humid, causing the washing-out of basic substances like lime from the soil minerals, and the progressive acidification of the soil. Earthworms, in particular, cannot tolerate acid conditions, and are absent, or nearly so, from coniferous-forest soils. Earthworms are the chief agents which mix dead plant residues with the mineral soil, and in their absence the dead and decaying forest litter does not get mixed with the soil, but, as already described in Chapter IX, accumulates as a peaty mass of raw humus on the surface.

The roots of the living plants are disposed in two quite distinct layers, the roots of the ground vegetation being mainly in the raw humus, and the roots of the trees in the layers overlying the hard, brown ' B ' horizon. (When soil conditions are unfavourable the main root systems of spruce trees may develop in the raw-humus layer, and the trees are then very liable to be blown over ; such conditions, however, are not typical of healthy soils.) Very few roots of any kind exist in the impoverished mineral soil between these two root-inhabited layers, one important result of which is that this soil has nothing in the way of a crumb or granular structure and, apart from its poverty in plant food, is a poor medium for plant growth.

A coniferous-forest vegetation consists of trees and other plants which can tolerate acid conditions. They can do so because they do not need much lime or other basic material (which is lacking in acid soils) in their make-up. Because they are poor in lime their needles and leaves produce an acid humus, which in the absence of earthworms appears as raw humus, unmixed with the mineral soil. The acid raw humus—itself a result of the shortage of lime in the soil, the nature of the vegetation and the absence of earthworms—produces humic acids which dissolve in rainwater and percolate downwards, strongly attacking the mineral soil particles. In time, practically everything except silica is washed out of the top horizon of mineral soil, what remains will not support plant life, and the horizon is almost devoid of plant roots. The material removed from this horizon is all deposited in a darker coloured horizon underneath, as shown in Plate 1a, but this is a chemical process independent of the nature of the living community. The fauna and flora, however, are the main agents responsible for the character and rather striking appearance of the upper soil horizon.

GRASS STEPPE

A steppe society consists of varying numbers of species, predominantly of grasses, but also of legumes and other flowering herbs, and a large faunal population which may include grazing animals as well as many burrowing animals, rodents, earthworms, insects, etc. There is a much larger faunal population, great and small, than there is in a coniferous-forest soil. The micro-organic population is also much larger, and contains proportionately more bacteria. The steppe climate is continental and rather dry—very cold winter and hot summer, most of the precipitation falling as snow.

The most striking difference between the steppe soil or chernozem (Plate 2a) and the coniferous-forest soil or podzol is in the distribution of humus, which in the podzol is concentrated mainly in the surface raw-humus horizon and in the chernozem is evenly distributed through the soil, sometimes to a depth of several feet. This even distribution of humus is brought about by two biological factors ; firstly by the ubiquity of the grass roots which interlace the soil in all directions and die and humify where they grew, and secondly by the activities of the fauna, not only earthworms, but also burrowing animals which are continuously mixing up the topsoil and subsoil. Burrows filled with darker soil can often be seen in the subsoil, and similarly burrows filled with lighter soil brought up from below are found in the dark topsoil.

The reaction of a typical steppe soil is neutral (pH 7), and in the semi-arid climate although there is some washing-out of lime, as may be seen by the occurrence of a lime-rich layer in the subsoil, it is, when once the soil has reached maturity, no more than can be compensated by the lime and other materials taken up by the plant and deposited in or on the soil when the plants die. Steppe and many other grassland plants are rich in lime, nitrogen and mineral nutrients, and large quantities get involved in the biological cycle and are stored in decaying plant and animal residues and in humus. As the soil's bacterial population and activity are high, organic decomposition proceeds rapidly during the growing season, and abundant nutrients are released for plant use. Thus the rich black chernozem has a deserved reputation for high agricultural fertility—provided it has enough water.

From an agricultural point of view the most important agents in the formation of a chernozem are grass roots. Grass roots are indispensable for the production of the stable crumb structure of a chernozem that always has been and probably always will be the goal of the cultivation of other soils. Bradfield (1937) has described how they work in these words :

' Practically all soils as far as I know are found to be in better physical condition after they have grown a heavy crop of grass or preferably a grass-legume mixture for a few years.

' For generations farmers have realised that for some reason cultivated crops grew better following such a mixture. Under grass most soils assume a granular or crumb structure. The exact mechanism of the formation of these crumbs is not fully understood, but the result obtained is universally regarded by soil physicists as the ideal physical state for the growth of most crops. It is best developed in soils which are saturated with lime and which contain from 3 to 10 per cent organic matter. Having these amounts of lime and organic matter in the soil, however, will not ensure a good granular structure. Something else is needed. The organic matter must be of the proper type and it must be properly distributed. I have tried to picture to myself how these granules might be formed and why they are so important.

Additional evidence is needed, but I feel that the picture is reasonably accurate.

' Grass roots are so numerous that in a well-established sod they are seldom over 3 to 5 millimeters apart. These roots ramify the soil in all directions. Each root represents a centre of water removal. As water is removed from the soil in contact with the root additional water moves toward the root by capillarity. As water is removed the small fragment of soil between the roots shrinks and is blocked off by the roots. The pressure developed by the capillary forces, compressing the granule from all sides, is great, in many soils it reaches over 5,000 pounds per square inch. As a result these granules become quite dense, their apparent specific gravity ranging from 1·8 to 2·0. The total pore space inside them is small and the size of the pores is very small. Water moves into them slowly but is held firmly. The pores are so narrow that they are easily completely sealed by capillary water, and as a result the ventilation of the interior is poor. Consequently, reducing conditions frequently exist in the interior of the granules simultaneously with oxidising conditions on their surface. This often causes a migration of substances which are more soluble when in the reduced form to the surface of the granule where they are oxidised and deposited. This deposit serves as a cement and helps to stabilise the granule. Very hard granules of soil may be formed by the compression due to surface tension alone, but such granules are not water-stable. When immersed in water, the air-water interface gradually disappears throughout and the granule breaks to pieces.

' In forcing its way through the soil many cells are sloughed off the living root and serve as food for bacteria. Eventually, the roots die and are decomposed *in situ*, forming a humified, often water-resistant coating around the granule. The marked difference in color between the surface of such granules and their interior is evidence of this. In many respects it seems to me these tiny granules may be compared

to miniature ' earths.' Most of the inhabitants live near the
surface where the ' air ' is better. In the strongly granulated
soil practically the entire mass of clay and silt particles are
clumped together in these water-stable aggregates. As a
result there are two fairly sharply defined groups of pores
in such soils, capillary pores within the granule and non-
capillary pores between the granules. The non-capillary
pores are relatively large. Water enters them readily but
is retained only at the periphery. This leaves a continuous
series of connecting chambers through which air can readily
pass. The water at the periphery is drawn into the capillary
pores between the unit particles making up the granules.
This water constitutes the most closely held reserve in the
soil. Such a soil has a permeability approaching that of
sandy soils combined with a storage capacity of the heavier
textured soils.

' Such are the structures which perennial grasses tend to
develop in soils ! Such soils provide optimum growing conditions for most crops. The organic matter constitutes a good
reserve of the elements essential for growth. Water and air
are present in the proper proportions. With an abundance
of food, water and air, soil micro-organisms flourish and
gradually convert the organic reserves into the simpler forms
required by crop plants.

' Crop roots can easily force their way through the large
well-aerated non-capillary pores. As a result, a more
efficient and extensive absorbing system is developed. The
food reserves are concentrated largely on the surfaces of
these pores. Once a soil is well granulated and when well-
aerated interfaces are formed, roots will tend to follow them
instead of forcing their way through the dense aggregates.
The effect of the roots is thus accumulative. It is not
definitely known how many seasons' growth is required to
produce the optimum structure. The major part of the work
is probably done in the first few years of growth of the
sod.'

The essential features of a chernozem, then, are its magnificent crumb structure, its high and evenly distributed humus content, its great biological activity and an ample supply of plant nutrients. There are other features, such as its depth, the presence in it of a layer of calcium carbonate, its neutral reaction and its aridity, that are attributable mainly to geological and climatic factors, but the features previously mentioned are the most important agriculturally, and are the result of the combined actions of the specific flora and fauna of a steppe. As with other soils, geology and climate exercise a general influence on the biologically determined soil properties because these factors influence the nature of the living association. In geological and climatic environments different from the steppe environment there are also grasslands, but with different associations of plants and animals, and different soils arise, showing in greater or less degree the typical characteristics of a chernozem.

Farm Land

It might be thought that there was no typical soil of farm land, but it might also be thought that a thinking pine tree, if it could sample existence in several podzol soils, would find very little in common between them, though thinking man would regard them as essentially the same type of soil. The pine would be interested mainly in the food and water supply in the immediate neighbourhood of its own roots—a localised and varying characteristic which the soil scientist would have to ignore. A farmer who had had experience of a fertile highly farmed clay and a fertile highly farmed sand would conclude that they could not rationally be regarded as the same type of soil, because they required very different management. But both display the most typical feature—high fertility—of soil which has been successfully worked by man for his own use. Whatever the physical nature of the terrain and whatever the climate, a farm soil can be defined for present purposes as one which will produce abundantly a few of the hundred-odd plants that man has domesticated.

The dominant biological influence in the evolution of farm soil is, obviously, man, working either alone to produce his own and his family's primary needs or, more frequently, as a member of some economic society to produce a portion of the needs of that society. Whichever rôle he plays he has to keep the land he works in trim or go under—sooner or later. He makes and pays for plenty of mistakes, and modern science, which has provided the knowledge by which he can now avoid some mistakes, has also provided opportunities for new mistakes. But slowly and inevitably lands in continuous human occupation have increased in fertility ; perhaps science's greatest contribution to agriculture will be to accelerate the process in recently settled lands.

As the trees in a forest or the grasses in a steppe produce the type of soil adapted to supplying their special needs, so man the farmer produces, or aims at producing, his particular type of soil. He does it mainly by his own labour—cultivating, manuring, draining and so on—but other members of the farm community, which could not survive without the farmer, play a part. The macrofauna of this community consists chiefly of grazing animals which, apart from their strictly economic functions, serve to keep grassland as grassland (at least in natural forest regions), and to provide excrement with which to decompose into manure the otherwise waste organic matter, like straw, that has been estimated to comprise about half of all the crops produced on a farm. Livestock thus enable to be returned to the soil nearly everything taken from it except that removed in human foodstuffs.

The meso- and micro-fauna and flora—earthworms, insects, bacteria and fungi—are, as far as we know, of the same general type as in a grassland soil, but the mesofauna tends to be less numerous in a farm soil. Some of the soil-forming functions of the mesofauna of grassland soil, such as mixing it up, aerating it, and making drainage channels, are performed in farm soils more expeditiously by cultivation. Where organic manure is used the mesofauna of cultivated soils consumes and excretes the manure, thus helping to humify it and incorporate it into the soil mass. Where organic manure is not used the mesofauna is numerically less, but that is not necessarily a loss to the farmer because there is less work for it to do. The farmer, if he is sufficiently clever, can find substitutes for both the humus and the plant nutrients in organic manure.

The macroflora of farm land is quite specific, though it varies in certain particulars from place to place. It consists almost entirely of plant species that are used directly or indirectly as human food, together with others like flax, cotton and tobacco, whose products are almost as necessary as food for the running of organised human society. The crop occupying by far the biggest area in the world's farm lands is grass. In Britain it occupied in 1950 about $18\frac{1}{2}$ million acres out of a total farm-land area of some 31 million acres, excluding nearly 13 million additional acres of ' rough grazing,' much of it uncultivable hill land on which man's influence has been confined to preventing the re-invasion of forest or the invasion of the Forestry Commission. After man himself, grass is the most important agent in producing typical farm-land soil. Although outside the typical chernozem environment, there is not usually the same favourable combination of climatic, geological and vegetational conditions for the formation of an ideal crumb structure, nevertheless wherever a mat of perennial-grass roots forms and grasses die and decay in the soil, some kind of crumb structure will be produced. One of the strongest arguments that can be advanced in favour of mixed farming is that it ensures that a part of the farm will be under grass, making or preserving a good soil structure. The argument becomes stronger for ley farming and ' carrying the plough round the farm,' because thereby all the land has a chance to recuperate its structure under a grass ley, and the restored fertility is then utilised in more productive arable cropping. But there is very little land in Britain that has not been in grass at some recent time (in 1939 there were only about 9 million acres under the plough).

The rest of the flora of farm land has a much smaller influence on soil properties, because it is composed mainly of annuals which only exert their influence for a few months, and their greatest influence (when fully grown) for a few weeks. The presence of growing plants in a soil, however, almost certainly has an effect on the microbial population, though not necessarily a useful effect. As mentioned in Chapter VI, the neighbourhood of plant roots is the chief seat of microbial activity. There is some evidence that each crop has its own type of rhizosphere population, but it is not known whether the rhizosphere population of, say, wheat has any agricultural virtues or vices not possessed by that of another crop. The one type of bacteria of which deliberate use is made in agriculture is the nitrogen-fixing type associated with legumes. The effect on soil productivity of growing legumes has been known from earliest times, and until, within living memory, ammonium and nitrate fertilisers became freely available, legume-nodule bacteria were the principal source of the additional nitrogen that agriculture must get from outside the soil to repair the inevitable nitrogen losses which cultivation brings in its train. Farm land should contain more nitrogen than non-farm land. It is conceivable that a practicable agricultural system may be devised in which all the necessary nitrogen is provided by fertilisers, but such a system is still a vision of the future. Wherever suitable legumes do not grow—as in many tropical countries—one of the necessary tasks of the reformer who wishes to inflict a settled life, health, prosperity and civilisation on the people is to find legumes that will grow. Nodule bacteria are therefore characteristic members of the microbial population of farm-land soils.

The pine forest makes a soil—a podzol—which is pre-eminently suitable for pine forest—chiefly because it is not suitable for any other plant association that might compete with coniferous forest. The steppe makes a soil—a chernozem —for grass and herbage plants, a soil with a beautiful crumb structure which ensures a high availability of plant nutrients and enables the vegetation to get all the nutriment it requires despite considerable shortage of soil moisture. Man makes an agricultural soil which, when he is successful, has a good crumb structure, is well drained yet water-retentive, and has ample plant nutrients (provided by manures and fertilisers) and moisture for abundant crop growth. Often enough, a farmer fails to make such a soil, but all the operations of good soil management are directed towards that end, and for the same reasons as the pine tree makes its poor podzol, because both man and pine will go under in the struggle for existence if they fail. The farmers of the last fifty centuries have not thought much about the basic principles of making an agricultural soil, indeed, only in the last half century have a very few farmers thought about them at all. Nevertheless, most of them have created, out of an infinity of materials, soils which carry the hallmark of man's influence in that they are more productive than Nature made them. In a country like Britain, and in more intensively farmed countries, most of the soils have been completely transformed into a new and more productive type, and in Britain almost every inch of land, at least in the lowlands, could within a few years be brought to a state of high agricultural productivity when the necessity arises—as it most surely will. The job will be expensive and difficult, but it can be done. One of the most costly operations will be the drainage of land, much of which was drained in the nineteenth century at a fraction of what it could cost now and subsequently neglected.

CHAPTER XI

SOIL CLASSIFICATION

UNLIKE most ' pure ' sciences soil science—if it merits the name of science, which some people dispute—has had to work for its living from its earliest days. Other branches of natural science were able to build up at least a foundation of theoretical principles from which practical applications sprang. Right from the beginning soil science was called upon to solve practical problems of agriculture before any basic principles had been formulated. The first soil scientists were chemists and geologists, then followed physicists and microbiologists, all applying the established principles of their sciences to the study of the soil. Only in Russia, and only for about thirty years, were soils studied as ' natural bodies,' as distinct entities in much the same way as plant ecologists regard forest and grassland communities. Now the Russians, like everybody else, have to direct all their attention to the practical application of science to increasing the productivity of the soil. There is no longer any time for, and no encouragement is given to, the study of soil as soil.

The Russians, led by V. V. Dokuchaev, were the first to elucidate the relationships between soils, climate and vegetation that have been very briefly touched on in Chapter I. The connection between climatic and vegetation types (e.g. that coniferous forest occurs in a cold moist climate, deciduous forest in a temperate moist climate and grassland in a semi-arid climate) was already known, and what Dokuchaev and his followers showed was that a similar connection holds between climatic and soil types. Since climate determines vegetation type it will almost certainly determine soil type also, since living things in general and vegetation in particular exert a predominant influence on soil formation. Dokuchaev drew up a classification of soils relating soil types primarily to the climates in which they were formed, and subsequently to the five main factors influencing their formation or genesis —climate, vegetation, geology, topography and time. This soil classification takes no account of a soil's suitability for agriculture or other utilisation by man. It can claim to be quite ' scientific,' but it cannot claim to be very useful for agricultural purposes.

In the vast flat central plain of European Russia, covered by glacial and aeolian deposits which vary comparatively little over wide areas, the influence of climate and vegetation on soil formation is exceptionally clear, because the other physical factors of geology and topography, which might have interfered with the expression of the climatic influence, are fairly constant. By contrast, in a small country like Britain where climate does not vary much (on a global scale) from one end to the other but geology and topography vary greatly and rapidly, these last two factors appear to have a much greater influence on the type of soil produced than has climate. We find a much greater difference between soils formed on granite and chalk in the south of England (different geology, same climate) than between two granite soils formed in the north and south of Britain, respectively (same geology, different climate).

GENETIC SOIL CLASSIFICATION

It is not necessary to describe the whole of the climatic system of soil types. The chief types important for temperate agriculture have already been referred to in Chapter I—the podzol, the coniferous-forest soil occurring in a cold humid climate ; the brown forest soil (deciduous forest) occurring in a temperate humid climate ; and the chernozem (grass steppe) occurring in a semi-arid continental climate. The types are defined by the morphological features of the soil profile or section—that is, by the number, thickness, humus content, chemical composition and structure of consecutive horizons in the soil profile. Note that at this level of classification no attention is paid to texture, which is nearly always agriculturally the most significant of soil properties. The main climatic types, in the order in which they would occur in the northern hemisphere passing from north to south over a flat plain as in Russia, are given in Table IV.

TABLE IV—THE CLIMATIC CLASSIFICATION OF SOIL TYPES

Soil type	Climate	Vegetation
Tundra soil (peat)	Cold or arctic	Tundra
Podzol	Cold humid	Coniferous forest
Brown (or grey) forest soil	Temperate humid	Deciduous forest
Chernozem	Sub-humid to semi-arid	Steppe
Chestnut-coloured soil	More arid	Steppe
Grey desert soil	Very arid	Desert or nearly so
Desert	No rain	Nothing
Red soil	Sub-tropical	Savanna or steppe
Lateritic soil	Humid tropical	Rain forest

A simplified climatic-soil map of Eastern Europe is shown in Fig. 4. It would do almost equally well as a natural-vegetation map by changing the soil-type names on the right of the map to the corresponding vegetation-type names given in the table above.

A similar climatic-soil map of Great Britain would show only two soil types, podzols and brown forest soils (and 'mountain soils' which are topographically, not climatically, determined), the boundary between the two corresponding with the boundary between the wet and dry halves of the island running from Bristol, east of the Pennines and along the east coast of Scotland.

The great regularity of the climatic-soil zones in Russia is connected with the fact that Russia is a vast plain in which the annual rainfall *decreases*, and the annual temperature

Fig. 4   Soil Map of Eastern Europe showing zones of genetic soil types

increases, from north to south. The decreasing rainfall means that less water reaches the soil, and the increasing temperature means that more evaporation takes place from the soil, going from north to south. We say that the *humidity* of the climate decreases from north to south (as far as the desert zone, after which it increases again), humidity being measured for this purpose by the excess of rainfall over evaporation. Where rainfall exceeds evaporation water will drain downwards through the soil, carrying soluble salts and plant nutrients with it ; where evaporation exceeds rainfall (aridity) water will be drawn upwards from the groundwater, again carrying dissolved salts and plant nutrients which will be deposited at or near the soil surface when the water evaporates. Consequently, the more humid the climate the poorer a soil will become in plant nutrients. Podzols are very acid and impoverished, brown forest soils are less so, chernozems, which form where rainfall and evaporation are about equal, are neutral and rich in plant nutrients, chestnut soils and grey desert soils become increasingly alkaline and sometimes contain so many harmful salts that little or nothing will grow in them even when they are watered (until the water has washed out the salts). South of the desert zone, which stretches in Eurasia from Russia to China in a broad zone around the 30th parallel of latitude, the climate becomes increasingly humid towards the equator, and the soils become more acid and poorer in plant nutrients.

A few other unnamed soil types, mainly mountain soils, are indicated in Fig. 4, which has been marked to indicate the *zonality* of the climatic soil types. The soil types occur in zones, corresponding to the climatic zones, and succeed each other from north to south. A similar succession of climatic soil zones is found in North America, but there the zones are more irregular and the succession tends to go from east to west—from podzols in the east to desert soils to the west of the Rocky Mountains. This is because the rainfall decreases as one goes westwards. In Australia, where the rainfall

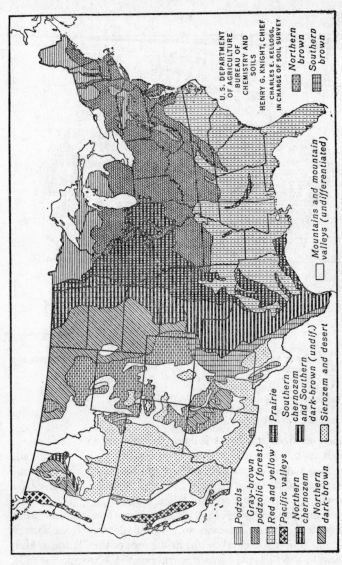

Fig. 5 Soil Map of the United States
(*U.S. Department of Agriculture*)

U. S. DEPARTMENT
OF AGRICULTURE
BUREAU OF
CHEMISTRY AND
SOILS
HENRY G. KNIGHT, CHIEF
CHARLES E. KELLOGG,
IN CHARGE OF SOIL SURVEY

Northern
brown

Southern
brown

Mountains and mountain
valleys (*undifferentiated*)

Podzols

Gray-brown
podzolic (forest)

Red and yellow

Pacific valleys

Northern
chernozem

Northern
dark-brown

Prairie

Southern
chernozem
and Southern
dark-brown (*undif.*)

Sierozem and desert

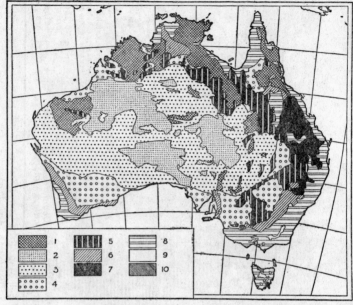

Fig. 6 Soil Map of Australia

1	Tablelands and ranges	6	Red brown earths and red earths
2	Desert sandhills	7	Black earths
3	Stony, loamy, and sandy deserts	8	Podzols and coastal marshes
4	Mallee soils and other brown soils of light texture	9	High moor
5	Grey and brown soils of heavy texture	10	Red loams

decreases from the coast inland, the succession of soil zones tends to be concentric. But since temperature and rainfall, the two main factors determining climatic humidity, do not increase and decrease together in North America and Australia the zonation is very irregular (Figs. 5 and 6).

A similar succession of soil zones is often found on mountains where temperature decreases and rainfall increases from bottom to top. There are chernozems at the foot of the Caucasus and, rising to the summit, zones are traversed of brown forest soils, podzols and mountain peats corresponding to the tundra soils of arctic regions. This is an example of ' vertical zonation ' of soil types.

A climatic soil type not yet mentioned is the *terra rossa* which is common in the Mediterranean region and in a few other places with a Mediterranean type of climate (mild, wet winter, dry summer). True *terra rossa* is only formed from limestone rocks, and is only found where both limestone and a Mediterranean climate occur together. It is a red soil supporting open forest vegetation, and is now usually found in a highly eroded condition, as most of the *terra-rossa* region has been under cultivation for two thousand years or more.

The *terra rossa* is an example of a soil type determined by both climate and geology. Climate is not always the determining factor in soil genesis. (The word genesis is used because the soil types we are discussing are often referred to as ' genetic ' soil types—types determined by soil genesis or the history of their creation.) Of the five genetic factors listed on p. 153, vegetation and climate usually act together, but geology and topography can in some circumstances so influence soil genesis that the normal climatic type is not produced. The types that are formed are known as ' intrazonal ' types, meaning that they are formed within a climatic zone when special circumstances prevent the formation of the normal climatic type. The presence of much lime in the parent rock is a frequent cause of the occurrence of intrazonal soil types. Black soils formed from hard limestone rocks in various climatic zones, including Britain, are known as ' rendzinas ' (a Polish word). Where there is sufficient lime in the parent rock to prevent the development of soil acidity brown forest soils occur as intrazonal soils within the podzol zone. Similarly, where the parent rock is very acid, as on the Bagshot sands in Berkshire and Hampshire, podzols occur within the brown-forest-soil zone.

Very striking intrazonal types are the saline and alkali soils, known by the Russian names ' solonchak ' and ' solonets,' that occur in low-lying situations in arid and semi-arid regions. Where the groundwater lies near the surface the intense evaporation causes the water to rise to the surface and evaporate, leaving behind any salts dissolved in the water. In dry regions the groundwaters are often heavily charged with sodium chloride and sulphate which, rising upwards, accumulate in the soil and may even form a white crystalline crust on the surface. The high soluble-salt content of solonchaks makes them quite infertile, and nothing except a few salt-tolerant plants will grow on them. The alkaline solonetses are formed by further evolution from solonchaks when the water table falls and the salts are washed out by rainwater, leaving sodium-saturated clay and humus in a highly dispersed condition, producing an unworkable slime when wet, and hard intractable clods when dry (Plate 2b). Solonetses are as unsuitable as solonchaks for agriculture, but both can be improved by appropriate treatment. Their formation is conditioned primarily by the depth of the water table ; in other words, topography is the determining genetic factor *within* the dry climatic zone. The solonets, however, is a soil type whose genesis, like that of all other soils but more obviously, is also conditioned by time, since it normally forms, by further evolution, from solonchak.

To complete the picture of genetic soil types reference must be made to the *azonal* soils which are not confined to any climatic zones. They are mostly very young or immature soils, and it might therefore be said that time, or lack of time, had been the predominant genetic factor in their formation. They include alluvial soils formed by continued deposition of mineral matter from rivers in their flood plains, fresh volcanic deposits, and ' skeletal ' soils composed of material like that of desert soils, high mountain soils and the residues of eroded former soils. Alluvial and volcanic soils often have great agricultural potentialities, but skeletal soils have little value.

We have referred to the soil types so far named in this chapter as genetic because their occurrence is conditioned by one or more of the genetic factors—climate, vegetation, geology, topography and time. When the types are arranged according to some kind of system, as in Table IV, we can speak of a genetic or, in this case, climatic classification of soils. The table shows how succeeding soil types are climatically related to each other : from the tundra to the desert zone the climate gets hotter and more arid, and from the desert zone to the equator hotter (roughly speaking) and more humid. By taking into consideration, as we must, the influence of other genetic factors on soil formation the classification system becomes more complicated because these factors (except vegetation which is the resultant of all the other factors) are not zonally distributed, but fluctuate irregularly from place to place. However, the genetic relationship between intra-zonal and zonal soils is usually quite clear. Thus the saline solonchak is a soil type of an arid climate that is formed under the influence of shallow and saline groundwater.

A genetic factor which is usually omitted in drawing up a genetic classification is man. Compared with the influence of other genetic factors on soil formation, that of man has usually operated for only a comparatively short time, and although man's influence can be revolutionary and can completely transform a soil type, the past and continuing influence of the other factors very largely determines man's correct management of the soil—which is what interests him more than does the academic question of soil classification.

The difficulty with soil classification is that a soil has no individual boundaries, as has a plant or an animal. We can isolate a typical soil sample, but we cannot isolate a typical soil. We cannot usually distinguish definite boundaries between two soils in the way we can distinguish the boundaries between one dog and another. In the soil maps shown in this chapter sharp boundaries are drawn between the soil types, but these are illusory and merely put in to conform to accepted ideas of map drawing. What happens is that one soil type usually gives way gradually to another, and there may be a wide transition zone between one well-defined soil type and its neighbour. The recognised types should be regarded merely as convenient points marked on a continuous and sometimes erratic curve of changing soil properties. The philosophy of soil classification has been well described by Robinson (1949).

The boundary between two soils can, however, be quite sharp, especially when geology or topography is the main factor determining the change of type. Thus, in England, the transition from rendzina to brown forest soil (both usually considerably modified by cultivation, but retaining some of their original characteristics) takes place quite abruptly on leaving the limestone country in which rendzina occurs.

The prime purpose of soil classification is to enable us to group the continuously variable entity soil (for no two soils, even only an inch apart, are identical in all respects) into a convenient number of classes or types, the approximate distribution of which can be shown on a map. Soils are classified so that soil types can be mapped. Soil types are mapped for a number of reasons, chief among which is that agricultural and cropping systems can with the help of soil maps be consciously planned to suit the soils.

For planning land utilisation the broad genetic soil classi-fication outlined above is not of much use because it tells us little that we could not have found out by studying more easily interpretable factors such as climate or geology. A chernozem, for example, is, according to its chemical, physical and biological properties, an apparently very fertile soil type, but it is *not* a soil type suited to intensive agriculture because it forms in a semi-arid climate with insufficient water. The most intensively farmed and most productive soils are marked on world soil maps as brown forest soils (in Western Europe). Chemically and physically the brown forest soils are not so fertile as the chernozems, and the cultivated Western European variety would be more correctly described as man-made soils, but very few scientists yet regard man as a natural factor of the environment, and the genetic system of classification is essentially 'natural,' like the very efficient modern classification of plants which is based on natural evolution. (But man has scarcely begun to interfere with plant evolution whereas he has been interfering with soil evolution for thousands of years.)

THE AMERICAN OR BINOMIAL CLASSIFICATION

For planning and advisory work which is the biggest use made of soil classification, much more detail is required than is given by the genetic system. We have to distinguish between soils in the same area that may be genetically similar, but differ in essential agricultural properties such as texture, nature of the parent rock (which will influence fertiliser requirements), drainage, soil depth, slope and so on. Many detailed systems of soil classification have been devised, most of them based ultimately on the genetic system, but emphasising in their finer subdivisions the agriculturally important rather than the genetic or historical features of the soil. These systems tend to have a national character because they have to fit into the broad outlines of national land-use policy, if any. Thus there is a Dutch system, a German system, a Russian system, a Hungarian system and several others on the continent of Europe alone. There is not a British system, perhaps because we have not had a land-use policy, but there is a British Soil Survey, which is fighting a determined battle against national indifference. The British Soil Survey uses the American system of soil classification, which with appropriate national modifications is also used by practically the whole English-speaking world.

In the American system soils are classified not, as in the genetic system, according to historical or environmental factors, but solely according to the visible characteristics of the soil in the field. Such a system is essentially practical ; soils which look alike and behave alike are classed together—as they should be in any classification. But very few soils look alike and behave alike, and the number of soil Types (written with a capital T to distinguish it from the totally different concept of genetic soil type) in the American system is legion. Soil Types are then grouped into successively larger classes on the basis of general similarities, the genetic types representing, as they must in any practical system of soil classification, a high stage or category. About thirty of them are recognised in all, and they are collectively called the Great Soil Groups (Category IV). They are further subdivided, according to differing features significant in land use, into (III) Families, (II) Series and (I) Types. Families (within one Great Soil Group) are distinguished from one another by differences in depth of soil and differences in the structure and consistence of the lower (B) horizon. Series (within one Family) are distinguished by differences in parent material and drainage conditions. Types (within one Series) are distinguished by differences in texture of the surface (A) horizon, and accompanying minor differences in the underlying (B and C) horizons. Thus the 'natural' Great Soil Groups are successively subdivided until in the Soil Type, which is the primary unit of detailed soil survey, most of the soil properties significant in land use are taken into account.

Soil Types differ only in texture within any one Series. Each Type is given a binomial designation, the first word representing the Series, which is usually named after the place where the Series was first encountered, and the second word indicating the texture. Thus Miami Loam and Miami Silt Loam are two American soil types identical in all respects except texture. The Soil Survey of Great Britain (1950) describes the soils of the Croxdale Series, occurring in Durham, as ' derived from Coal Measure Boulder Clay, with impeded drainage. . . . Typically they are shallow, grey-brown stoneless sandy loams overlying dull yellowish-brown sandy clay. Rusty mottlings, indicative of slow drainage, occur throughout the profile. With adequate artificial drainage this is a reasonably good agricultural soil.' Nowadays it is becoming common for soil surveyors to identify the Series and Type categories—that is, to have only one Type in a Series because it is found that differences between Types within an already recognised Series are often agriculturally more important than differences between Series. The Croxdale Sand and the Croxdale Clay (if they existed) would probably be agriculturally more diverse than the Croxdale Sand and the Sand member of some other Series. This would run counter to the purpose of classification.

When a farmer is told that part of his land consists of Croxdale Loam he is not much better informed than before, but he can get from the Soil Survey a comprehensive and even comprehensible description of the soil with details about its distribution, response to management and recommended treatment and use. Some hundreds of Soil Series and Types in Britain have already been described and named, and gradually a complete card index of all British Soil Types is being built up. For the farmer it will be a card index rather than a classification, but a card index is what he requires. There is nothing in the designations to show how the Croxdale Series in Durham is related to the Bodley Series in Devon, and to the Durham or Devon land user there is little of interest in the details of the other's soil. But to the soil surveyor further classification of Series into the higher categories of Families and Great Soil Groups does bring out genetic similarities and differences which enable him to give practical interpretation to the prosaic facts recorded in his soil descriptions.

When a soil surveyor has classified his soils he proceeds to put his findings on to a map. On small-scale maps, like Figs. 4–6, only high classification categories like Great Soil Groups are indicated. Such soil maps represent a kind of compromise between vegetation maps and climatic maps ; indeed, the positions of the (in reality) vague and diffuse soil boundaries are often determined by the less vague boundaries between vegetation types or by some arbitrary formula defining the climate by some relationship between annual rainfall and evaporation, i.e. by a measure of the humidity. Such maps would be helpful in deciding on the location of large-scale land-development schemes in undeveloped country.

In highly developed countries like Britain large-scale maps showing Soil Series or Soil Types are required for farm advisory work or detailed land planning. A Soil-Type map will show local variations in such properties as the humus and nutrient contents of the soil, drainage conditions, texture and perhaps state of cultivation. These variations are not, of course, marked as such on the map, but can be readily ascertained by referring to the descriptions of the Soil Series or Types that accompany the map. In Britain, local differences in soils are associated primarily with differences in surface geology and secondarily with topographical features which are usually related to geology. 'Detailed' soil maps are usually made in Britain on a scale of 6 inches, and 'reconnaissance' maps on a scale of $2\frac{1}{2}$ inches or 1 inch to the mile. An example of a soil-series map is shown in Fig. 7.

Several other classification categories besides those already mentioned have been used, combining Soil Series or Types into more comprehensive units, or splitting them up into smaller units. The Soil Survey of Scotland uses the Association, an originally Canadian concept, to include several Series derived from the same parent rock but having different drainage conditions caused mainly by topographical variations, and hence markedly different use capabilities. The separate Series are described as freely, imperfectly, poorly and very poorly drained members of the Association. The Soil Type or Series can also be further subdivided into Phases on the basis of slope, depth of soil, degree of erosion, stoniness and so on. These distinguishing properties may have great influence on the use that can be made of a soil. In the American system of classification the Phase is enumerated as the lowest category (0), a sub-division of the Type, but recently it has become the practice to subdivide any category, even a Great Soil Group, into Phases.

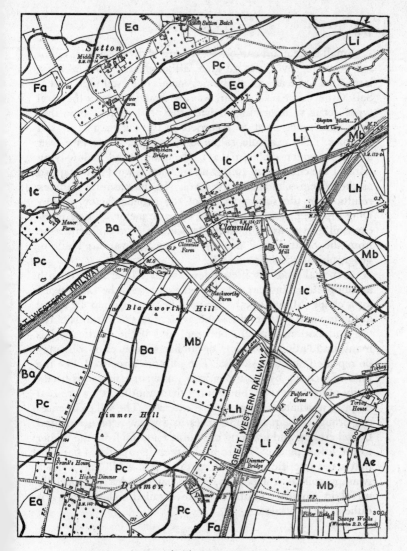

Ba Badsey Li Lydford Pc Podimore Fa Fladbury Ea Evesham

Mb Martock Lh Long Load Ic Isle Abbots Ae Atrim

Fig. 7 Part of the 6-inch to the mile soil map (reduced) of part of Somerset

(*Reproduced from the Ordnance Survey map with the sanction of the Controller of H.M. Stationery Office. Crown Copyright reserved.*)

Small-scale maps can obviously only show the predominant soil occurring over a large area, although it is unusual for a large area to be absolutely homogeneous in its soil type (Great Soil Group). Areas are frequently found where other soil types occur so intermingled with the predominant type that their occurrence must be recorded on the soil map as a 'soil complex.' For example, saline soils (solonchaks), formed over a shallow water table (p. 160), are often found in low-lying parts of the chernozem zone, and such areas would be mapped as chernozem-solonchak complexes. In this case the complex results from the superposition of the genetic factor of low-lying topography on the general climatic factor. The American Soil Survey uses the word Association (in a different sense from the Scottish Soil Survey) to describe various kinds of geographically associated soils.

A special kind of soil complex, associated particularly with an undulating landscape, is known as a *catena*, the Latin word for a chain (Milne, 1935). Quite different soils may be found at the top and bottom of a slope, with others or variants of the top and bottom types in between, and if the landscape has many undulations the same or similar sequences of soils may be repeated on other slopes. The soils are associated, like the links of a chain, by the topography, and since no soil is necessarily more typical of the country than another, they are best mapped as a catena, or as several catenas. The soil differences are brought about by differences in drainage conditions and by soil erosion and leaching of soluble material from the higher to the lower places. In tropical Africa it sometimes happens that a red lateritic soil occurs at the top of a hill and a black swampy soil at the bottom. The catena idea is especially useful in tropical countries where detailed soil surveying is difficult or impossible.

PRODUCTIVITY RATINGS

In the American and similar systems of soil classification the designation of soils, at the type or series level, is and must be purely descriptive. Classification proper begins when related soil Series are grouped into Families, and related Families into Great Soil Groups. This is the correct way of constructing a system which sets out to be both scientific (up to a point) and practical (up to a point), but the ordinary man may be excused for asking what it all means in terms of usefulness, and the ordinary soil surveyor has to spend quite a lot of time explaining it. The first thing the user of the Croxdale or any other Series wants to know is not the details given on p. 164, useful as they may be, but whether the soil is good or bad. He probably knew that long before the soil surveyor arrived, but he would still like to know whether the soil is better or worse than other soils, and by how much.

A quantitative picture of soil fertility is much easier to grasp the meaning of than a mere description. In order to satisfy the increasing demand for definite figures of soil fertility a number of systems have been devised for making quantitative estimates of productivity of soil Types from various kinds of data. Such estimates are known in America as 'productivity ratings,' and tables of productivity ratings of all the soil Types in the area surveyed are included in the soil-survey reports of the United States Department of Agriculture.

Productivity ratings are calculated from crop yields actually obtained on the soil under average and good systems of management. The productivity rating is 'the crystallised expression of the experience of the people who have used or are using the land' (Ableiter, 1940). Ratings are given to specific soil Types as defined by the U.S. Soil Survey. In the soil-survey reports, tables are given of productivity ratings or indices in relation to all the main crops of the district, e.g.

Soil Type	Corn		Wheat		Rye		Alfalfa		Sugar beet	
	A	B	A	B	A	B	A	B	A	B
Miami loam	70	90	70	100	80	100	70	90	60	70

The figures under A and B refer respectively to the percentage of the standard yield of the crop obtained with (A) common practices of management (average farming) in the area, and (B) the best current practices. The standard yields are selected to represent the approximate average yield obtained for that crop on the more extensive and widely developed soils of the regions in the United States in which the crop is a principal product. The standards refer to average yields obtained without the use of fertilisers. The standard for corn is 50, and for wheat 25 bushels per acre ; for clover and timothy it is 2, for alfalfa (lucerne) 4 tons. Under average management for the county in question the Miami loam will therefore yield 35 bushels of corn and $17\frac{1}{2}$ bushels of wheat, and under first-rate management 45 bushels of corn and 25 bushels of wheat. The difference between the ratings is a measure for the soil's response to good management.

Several other systems of evaluating productivity are based not on actual yields obtained from the soils, but on selected soil properties such as texture, depth, humus content, drainage, acidity, etc. to which ' marks ' are assigned and the productivity calculated by an arbitrary formula. These systems were much favoured in Germany before the war, and similar systems have been proposed in America (Storie, 1933) and England (Clarke, 1951). They are of necessity extremely subjective and dependent on the judgment of the observer, but in the hands of expert users they appear capable of reflecting relative soil productivities with reasonable accuracy. If, however, anything happens to the expert user the system and the formula are liable to break down.

Storie's system may be taken to illustrate these inductive methods of rating soils for agriculture. In this system the ratings are calculated from percentage values awarded to certain soil characteristics. Three ' factors,' A, B and C, are used, one referring to the general character (excluding texture) of the soil profile and particularly to stratification and degree of weathering, another to the surface texture, and the third to ' soil-modifying conditions ' such as drainage, acidity and alkalinity, erosion, etc. The values of the three factors, expressed as percentages of the optimal conditions for plant growth, are multiplied together to obtain the rating, and the product is expressed as a percentage of the maximum.

$$\text{The rating } R = \frac{A}{100} \times \frac{B}{100} \times \frac{C}{100} \times 100 = \frac{A \times B \times C}{10,000}$$

The advantage of multiplying, instead of adding, as is done in the German systems, the marks credited to each soil characteristic is that thereby any one abnormal factor can dominate the final rating. A soil may have excellent profile characteristics (A) rated at 100, excellent texture (B) rated at 100, but bad drainage (C) rated at 10 that will spoil the soil for agriculture. The product A×B×C gives a percentage rating for the soil of 10, according with the fact that despite other favourable conditions the bad drainage renders the soil unproductive. Adding the marks would have given 210 out of a possible total of 300 or a percentage rating of 70. On the other hand, for a soil for which all three factors were rated at 50 (and therefore one would expect the rating for the soil to be 50), multiplication of the factors gives a rating of only 12·5. All inductive productivity-rating systems suffer from the weakness that the formula used for integrating the separate ' marks ' is not valid over the whole range of productivity.

LAND CLASSIFICATION

The two main purposes of soil classification and subsequent soil mapping are to show farmers and foresters how to make the best use of their land, and to help local and national administrators to plan the broad picture of desirable land use. For the first a large-scale soil map is required, and should be able to answer most of the questions' that can fairly be asked of it. For the second neither a large-scale nor a small-scale soil map suffices, though it can help. Here it is not so much a question of whether the soil is suitable for permanent pasture or arable as whether the land can be used best in the public interest for agriculture, forestry, building, aerodromes, recreation, etc. Many other factors besides soil are involved—marketing facilities, social amenities, military requirements and present and future developments in the local or national economy. All these factors are included in the economic concept of *land*, as distinct from soil, and to satisfy the increasing demands of planners numerous systems of land classification have been drawn up.

Land classification is a combination of a classification of the physical environment (soil), which can be described objectively, with a classification of the economic and social environment which, by its very nature, can only be described subjectively. On account of this large, highly subjective element, the first essential is to define precisely the purpose for which the classification is to be used. The clearer the definition of purpose the clearer will be the classification of land. A vague purpose like ' using land for the greatest good of the greatest number ' gets nowhere. The United States Soil Conservation Service uses a simple land-classification scheme whereby land is put into one of eight classes according to the extent of the special measures required to maintain fertility and prevent soil erosion. The term *land* is here used in a restricted sense—as something which may, but should not, erode—and the classification is made solely on a physical basis of soil, climate and topography. It is simple because it ignores social and economic factors which cannot usually be ignored in planning land use.

If we confine attention to agricultural-land classification the definition of purpose can be simplified in most cases to the maintenance or enhancement of soil fertility by utilising the land in ways that will best achieve that purpose. For if land is not so utilised it will be liable to deteriorate, and nobody deliberately plans land deterioration.

In a country like Britain, with a long agricultural history, a practical classification of land is according to present use which to a large extent reflects ' the combined influence of the varied factors concerned : the physical factors of relief, soil and climate ; the more purely economic and social factors of markets, prices, transport, labour ; the historical factors of land ownership, local usage, tenure, and even tradition and custom ' (Stamp, 1948). Obviously, however, when large areas have recently changed their use, as has happened in the last decade under the stress of war, the influence of past historical factors may be masked, but it has been found in Britain that the best and worst classes of land have mostly retained their pre-war use, and that the biggest changes have occurred on the intermediate classes.

As soil surveying in Britain is only now getting into its stride, it has been impossible to use soil types as a basis for land classification, but before the last war the Land Utilisation Survey of England and Wales under the direction of Dr L. Dudley Stamp carried out a very detailed land-utilisation survey of the country. Stamp (1948) used the survey to classify the country into the following ten classes of land distinguished, with reference to their economic productive value, according to site (elevation, slope and aspect) and soil (depth, water conditions and texture). The figures show the estimated percentage area of England and Wales occupied by each class.

Major Category I—good   48·8

1	First-class	5·5
2	Good general-purpose farmland—	
	2(A) Suitable for ploughing	20·3
	2(AG) Crops or grass	6·3
3	First-class, restricted use unsuitable for ploughing	3·6
4	Good but heavy land	13·1

Major Category II—medium   31·4

5	Medium light land—	
	5(A) Suitable for ploughing	5·9
	5(G) Unsuitable for ploughing	0·8
6	Medium general-purpose farmland	24·7

Major Category III—poor   16·1

7	Poor heavy land	2·1
8	Poor mountain and moorland	11·5
9	Poor light land	2·1
10	Poorest land	0·4

Residue—closely built over   3·7

Although this classification will doubtless be improved and modified it gives a useful idea of the proportions of good, bad and indifferent land we have at our disposal.

Plate 8   Severe erosion caused by overgrazing and lack of fertilisers.   The land on the left of the photograph has been
properly managed

Photo : Tennessee Valley Authority

Plate 9 Gullies formed by soil erosion in Stewart County, Georgia

Photo : Soil Conservation Service, U.S. Department of Agriculture

CHAPTER XII

SOIL EROSION

det'

SOIL erosion—the partial or complete removal by rain or wind of the earth's thin cover of living soil from which all dry-land life springs—is a phenomenon which within the last half century has attained the proportions of a world-wide epidemic disease. The areas that have been affected by soil erosion are enormous. An erosion survey of the United States, made in 1934, indicated that nearly two-thirds of the total land area had been damaged (Lowdermilk, 1935) ; for New Zealand the area estimated to be affected by soil erosion is two-fifths of the total, or more than two-thirds of the occupied land surface (Cumberland, 1944). Estimates for other countries *affects nearly all countries* have not been made, but erosion has occurred on a comparable scale in Australia, South Africa, tropical Africa, the West Indies, parts of South America, India, China, Russia and the Mediterranean. Indeed, about the only part of the so-called civilised world that has not been seriously damaged by this modern disease is western Europe, the most densely populated region of its size in the world. Perhaps 3,000,000 square miles of agricultural and pastoral land have suffered more or less severely from erosion—but that figure cannot claim to be much more than a guess.

THE CAUSES

The commonest cause of this colossal destruction of soil has been a decline in soil fertility and, more immediately, the loss of soil structure that accompanies a decline in fertility. Loss of soil fertility has been brought about by the adoption, especially in recently settled lands, of agricultural and pastoral practices unsuited to the soils and climates. Such maladjustment was inevitable. All that the early settler could hope for was to keep himself alive and earn a living, and he used the tools available for that purpose in the way his knowledge and common sense dictated. It may be assumed that soil erosion is a normal phenomenon following the opening-up of new land to agriculture, and possibly also of the wearing-out of old agricultural land. Exceptionally, it may not have happened on a large scale in western Europe because of the uniquely moderate climate, and especially the almost complete absence of prolonged droughts, torrential rains and high winds. The European settlers, therefore, who colonised most of the New World had no reason to take thought of the need of protecting the soils from rain and wind. They entered the new lands without any knowledge of the natural phenomenon of soil erosion, because they had never encountered it.

As it happened, in the sixteenth to nineteenth centuries practically the whole world felt the impact of western European culture. Nearly every important advance in technique and practice, including agricultural, has stemmed from western European stock. Had it not been for western Europeans, North and South Americans, Asians, Australasians and Africans might still be harmlessly scratching the soil with a hoe—as many of them still do. There would then have been no soil erosion to speak of. Europeans, therefore, are responsible, but not to blame, for most of present-day soil erosion. Posterity will possibly thank them for what they did. Soil erosion is teaching their descendants an extremely effective and healthy lesson in readjustment. The lesson is, generally, being learnt, and it may cut down by a long time the period that would otherwise have been required to establish a new and fruitful balance with Nature.

That soil erosion is caused by misuse rather than overuse of land is shown by the fact that in many (but not all) of the most densely populated regions erosion is not a serious menace, whereas it is a very serious one in quite thinly populated regions like North America, Australia and South Africa. The three outstanding types of misuse that in one form or another have been the immediate cause of most of the recent soil erosion are excessive deforestation, overgrazing of grassland and monoculture. These are merely three ways of depriving a soil of its plant cover.

Deforestation → leads to erosion

most important

Forests provide the most perfect anti-erosion protection that any type of plant association gives to the soil. The forest canopy holds back much of the rain and deadens the impact of what does fall on to the soil. More important still, in many forests a layer of decaying litter covers the soil and acts as a huge sponge which absorbs all the water that falls on to it and releases it slowly and harmlessly to the underlying soil. When this protective cover is destroyed over wide areas erosion happens rapidly.

Forests serve another purpose, connected with their soil-protecting function, that is almost equally important. They are the natural cover over most of the high lands where streams and watercourses start. The great water-holding power of forest-covered soils ensures that an even and regular supply of water reaches the streams from the soils, but when the soils have gone there is little to regulate the flow of water over the surface of the land. Streams then tend to get overfull in wet periods and to dry up in dry periods. The same effect, reproduced in hundreds of little streams, has a great cumulative effect on the flow of water in the larger rivers fed by the streams, and often causes serious flooding, destruction of irrigation and hydro-electric works and dislocation of the natural drainage system in places far removed from the deforested area. Some of the worst floods of recent times have started from deforested hill land. Nowadays there is a growing recognition of the importance of keeping watershed areas under forest as the most effective way of ensuring a regulated and controllable water supply to catchment areas.

Overgrazing

Overgrazing, though usually less spectacular in its results than deforestation, is much more widespread and has caused much more erosion than deforestation (Plate 8). In arid and semi-arid regions—and much of the world's natural grassland comes into these categories, since humid areas are forested—over-grazing is almost universal. Semi-arid regions are also the ones most liable to soil erosion. Seasons are variable, the grazier knows that he must expect a serious drought once in so many years, and there is almost invariably a tendency to stock up to the limit in moist years, which means that the pastures are overstocked and consequently overgrazed in sub-sequent drought years. Overgrazing means destruction of the vegetation and exposure of the soil, and also destruction of soil structure by cattle trampling the soil and thus increasing liability to erosion. Sheep and goats are the worst offenders. Goats usually get most of the blame, but they may merely give the *coup de grâce* to what sheep have already done.

Monoculture

It is not monoculture as such—that is, the cultivation of the same crop year after year—that causes soil exhaustion and erosion, but the absence from monocultural systems of alter-native crops that will give a soil protection from erosion and a chance to recover its fertility. Sugarcane, which gives a soil quite good direct protection and produces a stable crumb structure with its roots, can be grown continuously, apparently for centuries, without causing excessive erosion. Most other monocultured crops are not so suitable for the practice, and the soil requires periodic rests from their cultivation under soil-improving crops. The most important resting crop, in-deed the most important of all crops in the world, is grass. Grass has almost miraculous properties of protecting soils from erosion and building up their fertility. To 'rest' a soil is not to fallow it (that is more likely to kill it), but to make it produce as much as it possibly can to feed itself with.

Soils are in some ways very like human beings or any other living thing. They can be, and have been, killed by overwork, but the best way to keep them alive and healthy is to feed them well and work them hard.

SOIL EROSION AND SOIL FERTILITY

Erosion is a process that is going on everywhere and always. It is the process that has shaped the Earth into hills and valleys and plains. In places where life cannot exist, such as high mountain slopes or arid deserts, erosion goes on as quickly as rock weathering—that is to say, as soon as a particle of rock small enough to be moved by water or wind breaks off by weathering from a larger bit of rock it is subject to erosion. That is the reason why bare mountain screes consist almost entirely of boulders and coarse bits of rock ; all the finer particles which might have formed soil have long since been washed down to the bottom of the scree where probably a soil has formed. Similarly, deserts consist mainly of coarse sand because all the finer material has been blown away. The fine wind-blown material from the deserts ultimately gets deposited as loess from which some of the richest and deepest soils in the world have been formed.

These facts indicate that a mass of small, loose, discrete particles, such as soils are largely composed of, is an unstable deposit on the Earth's surface, though soils themselves in their normal condition are stable enough. The particles are continually on the move until they come to a place like a sheltered valley where they may be protected from the action of running water or of wind. How, then, have the masses of finely divided rock material that comprise soils accumulated over most of the dry land of the Earth ? The answer is to be found in the biological nature of soil and soil fertility, and it also supplies the answer to why soil erosion occurs and how it can be prevented.

The fact that soil material is unaffected by the normal processes of erosion shows that it must possess some property or properties not possessed by the mass of mineral particles of which it is composed. These properties are those associated with fertility and with the product of fertility—plant and animal life. The most important property which prevents erosion of a fertile soil is its structure which gives porosity and permeability to what would otherwise tend to settle down to an impervious mass of rock particles off which water would run, carrying particles with it. In a porous structured soil water is absorbed and held by soil and thus rendered non-erosive. Different soil structures vary in the resistance they offer to erosion. One of the most resistant is the crumb type, produced by growing grasses and generally associated with a high level of agricultural soil fertility. When the fertility of such a structured soil declines and the biological forces creating and holding together the soil crumbs weaken, the soil begins to revert to the condition of a loose mass of rock particles, in which condition its erosion is natural and inevitable.

The aggregates in a fertile soil are not the only factor producing soil stability. An equally important factor is the vegetation the fertility produces and which acts by affording the soil surface direct protection against the beating action of rain and the tearing action of wind. These two characteristics of a fertile soil—structure and a plant cover—are complementary in preventing soil from eroding. In a forest the protection afforded by the living plants is complete, and we find that forest soils often do not have, and do not require, a well-developed erosion-resisting structure. A forest podzol is almost structureless, but quite stable under undisturbed forest. The direct protection given by a grass cover is less than that given by a forest, and to compensate for this grass-land soil develops an erosion-resistant structure. One result of this is that some forest soils are liable to erode immediately after deforestation, whereas many years may elapse before the structure of newly broken grassland is destroyed and the soil begins to erode. It is possibly for this reason that deforestation has captured the public imagination as the main cause of soil erosion. The effect often follows dramatically after the cause. On the other hand, several decades passed between the time when the American prairies were first broken by the plough and the time when the ' Dust Bowl ' thrust itself on people's attention. In the interval continuous cultivation had broken down the structure of the soil to a powder which was easily blown away.

The Effects of Soil Erosion

Soil erosion is liable to occur whenever the balance is disturbed between the physical forces of running water and wind that tend to keep soil particles continually on the move, and the mainly biological forces that check the movement. These biological forces—the protective action of living vegetation and the structure-forming action of living and decaying vegetation—are produced by living agencies, consequently if a soil loses its capacity for producing living matter, if it loses its fertility, it will become susceptible to erosion. That is what does, in fact, happen. The most common cause of soil erosion is a decline in soil fertility, and the best way of preventing it is to increase soil fertility.

As fertility declines the soil gradually loses its structure, the humus which binds the separate particles together disappears, the soil loses its porosity and its capacity to absorb and hold water, so that water falling on it runs off the surface carrying the loose soil particles with it. At first the process may be slow and imperceptible. Perhaps a fraction of an inch of soil will be removed in a year. But the next exposed layer will be more erodible because the fertility and structure of a soil decrease with depth, and it will go more readily. So the process goes on with constant acceleration. When once the soil has gone and the material beneath is exposed to erosive forces we get an illustration of the extraordinary instability of loose rock material as compared with soils. Even quite a small stream of water can tear a gaping hole or gully, like that shown in Plate 9, in the land in an incredibly short time. Enormous tracts of land have been ruined by the formation of such gullies.

The effects of erosion are not confined to loss of productive and frequently irreplaceable soil, nor to the place where erosion occurs. Loss of the absorbent surface soil causes profound disturbance to the normal regime of groundwaters throughout the entire drainage system of the region. During the wet season an abnormally high proportion of the rainfall runs straight off the valley slopes into streams and rivers, causing floods and silting-up of the river beds with eroded soil carried in suspension. During the dry season the flow of water is abnormally reduced and may cease altogether, since what remains of the surface soil cannot store enough water to feed the rivers. For the same reason the permanent water table sinks and may cause the drying-up of wells and watering-places. The whole tendency in an eroding region is for the humidity effective in promoting plant growth to be reduced —for the country to become desiccated and to approach desert conditions even though the rainfall remains unchanged. Droughts in South Africa are much more severe today than they were fifty years ago—not because less rain falls, but because there is much less soil to hold the rain until crops need it.

The precipitate run-off of water which should be stored by the absorbent soil may constitute a more serious loss and have more far-reaching consequences than the loss of the soil itself. Erosion promotes run-off by removing the absorbent soil, and run-off accelerates erosion, since it is the principal agent in actually removing the soil. Water containing eroded soil in suspension has a much greater abrasive action than clear water. Erosion, once started, accelerates of its own momentum, and the further it advances the more difficult it becomes to control. Where land is required for agriculture, control is essentially a matter of giving the soil a stable physical structure and the

Plate 10 Terracing and strip-cropping on a Texas farm

Photo : *Soil Conservation Service, U.S. Department of Agriculture*

power to stand up to erosive forces even when deprived of the protection of vegetation.

The basic cause of wind erosion is the same as that of water erosion, namely, loss of soil fertility and the accompanying deterioration of soil structure. The porous water-holding crumbs of a fertile soil are not easily moved by wind, but the separate particles into which the crumbs break down are easily moved. A plant cover which breaks the force of wind is the best protection, and where wide exposed areas have been deprived of their natural plant cover by cultivation or over-grazing wind erosion may have most dramatic effects. Large areas may be swept clean of soil in a single ' dust storm.' The wind-blown soil is sorted out into particles of different sizes, the coarser sand particles being carried short distances and deposited in drifts against houses, fences, etc., while the finer particles may be carried for hundreds of miles in the form of a dust cloud.

Soil Erosion and Society

There have been periods in the past when soil erosion must have occurred on an extensive scale, at least in relation to the size of the civilised world. Soil erosion caused by deforestation round the headwaters of the Euphrates in Asia Minor played a part in causing the silting-up of irrigation channels and the floods that destroyed the irrigation system of Mesopotamia. The decline of the Roman Empire was accompanied by wide-spread erosion caused by deforestation and a general fall in farming standards and soil fertility. Deforestation of Palestine has continued through the centuries from Roman times, and was completed by the Turks in the First World War. Goats in their thousands, eating everything that appeared above ground, took the place of trees on the hills of Judea, and the unprotected soil was washed off the hills which are now almost barren. Wind-eroded soil has buried cities in Syria and in North Africa.

These are all examples of erosion taking place in the twilight of a country's history, as may be happening today in Eastern Asia. Whether erosion caused the decline of the country or the decline of a country and of its agriculture and soil fertility caused the erosion is not always clear, but history shows that the two have often gone together. But most present-day soil erosion, is taking place in countries where organised human society has been functioning for quite short times, and in none of these countries, particularly the United States and British Dominions, has human society yet reached its full development or shown any signs of going into premature decline. As a social phenomenon—and that is what soil erosion essentially is—it is quite different in new countries from what it is and has been in old countries. In new countries it is symptomatic of a stage in which society has not yet adapted itself to its physical environment. In old countries it indicates that society has outworn its environment ; perhaps the economy has been unable to keep pace with time, and population has expanded beyond the combined powers of agriculture and industry to support it, soil fertility begins to fall, reacting upon the vigour of the people, and a vicious circle is entered from which at present escape is difficult, if not impossible.

Although there are no details available about the early stages of development of former societies, it is probable that soil erosion, following on the destruction of the natural plant cover and before agriculture had developed an equally good protection for the soil, often occurred. There are, for example, sure signs of erosion on the hills of Wales that probably occurred when first the hills were deforested. We can identify the present topsoil with the subsoil of a former forest podzol, and sometimes deposits of the former topsoil can be found in the valley bottoms. The present soils and vegetation (grass, moss, heather, etc.) have re-established a stable equilibrium and, as can be seen from the clearness of many Welsh streams and rivers, little or no soil erosion is taking place.

At the present time some slight but quite noticeable soil

erosion is going on in parts of Britain, particularly in the fen land of East Anglia and on light soils in Yorkshire. Quite dark dust storms may sometimes be seen in the fenland after a dry spell in the spring when the light peaty soil has just been sown, but the crops have not yet appeared. This erosion causes some loss to the farmers, but chiefly in the labour of having to resow their fields. It cannot in any way be considered a national menace, as can erosion in other countries.

Man does not destroy the land he has conquered from Nature out of greed or folly, but owing to social and economic maladjustment. Social readjustment is not primarily the business of the soil scientist, but rather of millions of individuals seeking to find their feet in new and often rapidly changing circumstances. Soil scientists have, however, learnt enough about soil in the last quarter century to give a physical definition of soil fertility, and have found that a fertile soil—one that will give high yields of agricultural crops—has just those physical properties which will best enable it to withstand the erosive action of wind and water. Thus the way to stop soil erosion is to build up soil fertility.

Building up soil fertility is also coming to be the best way for a farmer to enrich himself. But it was not always so. In new lands the most economic, and often the only economically feasible, system of agriculture is to draw on the fertility stored up in the soil by centuries of wild-plant growth. It was by such a system that North America, for example, was colonised. Men cleared the land of the virgin forest or prairie, skimmed off the cream of soil fertility and moved on when the land showed signs of becoming exhausted. A similar system of ' shifting cultivation ' is the basis of all primitive agricultures, in which exhausted land is abandoned to the natural vegetation which in time restores its fertility. In much modern settlement exhausted land has been merely handed on from one owner to another, who has continued the process of exhaustion until the soil has washed or blown away from under his feet. One of the chief causes of soil erosion in tropical countries has been the shortening of the fallow period, as a result of increasing pressure of population on the land, in native shifting cultivation so that the soil has not enough time to recuperate before it is again cultivated, and gradually becomes exhausted and erodes away. Under native systems (say, 1-2 years' cultivation, 25 years' fallow), soil deterioration seemed not to occur, but a system of land utilisation that uses only 4-8 per cent of the available area is impracticable under any modern civilised economy. In these cases the systems of land tenure, evolved to meet the then prevailing economic conditions, are at variance with present conditions. To change those conditions is usually a matter of gradual social evolution, though it can be done, as in the Soviet Union, by revolution. But in the Soviet Union it was not done to stop, and it has not stopped, soil erosion.

Soil Conservation

Terracing

The basic principle of preventing soil erosion by water can be simply stated. It is to stop water running downhill on sloping land. One way of doing this is to break up the slope into a number of level terraces. Terracing has been used as an anti-erosion measure from the beginnings of agriculture, and in one form or another is very widely used today. There are flat bench terraces, broad-based terraces, contour furrows—level ditches run along the contours of the land to catch and hold run-off water—and various types of ridging and bunding. A much favoured practice is to plough along the contour, thus making a series of level ridges and furrows that will hold water, whereas the traditional European up-and-down ploughing creates incipient gullies down which water will pour, carrying soil with it. Always the aim is to stop water moving over the soil surface and to get it to soak into the soil.

In many countries where soil-erosion control is being attempted this is as far as matters have gone. Measures which check the movement of run-off water are quite effective, but they do nothing to remove the root cause of soil erosion, which is loss of soil fertility and the breakdown of erosion resistance which is a property of soil, but not of the rock particles of which soil is composed. At best these mechanical measures will stop further erosion from occurring, but by themselves they do not constitute a complete system of erosion control.

Strip-cropping

In the more advanced countries increasing attention is being paid to methods of erosion control or, to give it a more comprehensive name, soil conservation of the same kind as those used with complete effectiveness by Nature—namely, soil protection and soil building by plants.

Some agricultural crops give practically no protection against erosion, for example, widely spaced crops like maize, cotton, roots and tobacco ; others like grasses and close-growing plants give excellent protection ; the main arable crops give only moderate protection. All annual crops have the disadvantage that they do not cover the ground all the year. Protection against erosion on arable land can be obtained by employing rotations which include several years under grass. The grasses not only directly protect the soil against rain or wind, but also help to build up soil fertility and an erosion-resistant structure, so that the soil is less liable to erode when next it is put under the plough. After a few years the grassed fields are cultivated and the cultivated fields are grassed. By combining such rotations with contour cultivation in the system known as contour strip-cropping (Plate 10) very good erosion control can be achieved under a wide variety of conditions. The system is capable of almost infinite variation. The nature of the rotation and the width of the strips can be adjusted to suit local conditions of slope and weather ; and a further advantage is that it encourages and almost compels the farmer to adopt diversified farming, the absence of which in the New World has been a main cause for the

decline in fertility and loss of soil structure which precede the
onset of erosion.

The importance of grass as a rotation crop in systems of
conservation agriculture can hardly be overestimated.   It is
being increasingly appreciated even in countries like Britain
where erosion is not a problem, because of its beneficial effect
in maintaining soil fertility.   Most systems of soil-conserving
agriculture include the growing of grass as the one indispensable
crop.   For grass is Nature's agent in producing a granular soil,
and we know no substitute for it.   Martin (1944) in Uganda
has shown how the use of a temporary grass crop can also
restore fertility to worn-out soils in the humid tropics.   By
rotating elephant grass (*Pennisetum purpureum*) with cotton in
contoured strips he got considerably increased yields of cotton,
and the soil did not erode, as it did when cotton was grown
continuously.   Martin showed that the grass improved the
physical structure of the soil, and it was this that made the
soil resistant to erosion.   Furthermore, he showed that dung
or green manures had no effect on soil structure and only an
evanescent effect, if any, on soil fertility.   Apparently the
active agent in restoring soil fertility is the grass roots which
mechanically split up and compress the soil into crumbs.

Strip-cropping can also be used against wind erosion.
The strips are then laid across the direction of the prevailing
wind, so that any soil blown from the bare strips is caught
and held by the grassed strips.   Wind erosion is in general
more difficult to control than water erosion, because it is
easier to stop water from running downhill than to stop
wind from blowing across an open plain.   To prevent wind
erosion, reliance has to be placed principally on biological
control—that is, on the maintenance of soil fertility and an
erosion-resistant physical structure.   A certain amount of
control can be achieved mechanically by means of windbreaks, and by cultivating so that the soil is left with an
uneven surface and in a cloddy condition, but windbreaks
give only very localised protection, and clods made by a

cultivating implement are liable to break down quickly unless the fertility of the soil is maintained.

Stubble Mulching

A measure which is gaining increasing popularity in Canada and the United States as a protection against wind erosion is known variously as stubble mulching, trash farming or plough-less fallow. In many regions of insufficient moisture it is still impossible to avoid fallowing, often every alternate year, so as to store up two years' rainfall for one crop. A ploughed bare fallow is a source of great danger from wind erosion. In stubble mulching the land is cultivated after harvest with special implements which cut the crop and weed roots a few inches below the surface, but leave the plants standing in the soil. The standing plants give good protection against wind erosion. Ultimately they fall on to the soil and decay, forming an organic surface mulch which can be parted in bands for sowing the next crop. The method, of course, is not without its drawbacks ; in particular it does not always kill weeds effectively, and weeds are great consumers of precious water. But it is very effective against wind erosion.

Gully Control

The treatment of gully erosion is largely a matter of land engineering. Little can be done with large gullies except to prevent them growing larger. The aim is to prevent the head of a gully from eating farther back into the land. The head can be revetted or otherwise strengthened, and the inflow of water which is causing the gullying can be controlled or diverted. Sometimes it is possible to make a virtue of a necessity by damming the gully and turning it into a water reservoir. Gully sides are graded, if necessary, and seeded or planted with quick-growing vegetation which will stabilise the soil. In the United States two fodder legumes, kudzu (*Pueraria thumbergiana*) and lespedeza (*L. sericea*), have been very success-fully used for stabilising gullies as well as exposed road and

railway cuttings which are very liable to erode unless rapidly
covered with vegetation. The difficulty is that vegetation
often will not grow rapidly on the non-soil material exposed
in a cutting or gully side. Preliminary tests with the soil
conditioner, krilium (p. 108), have given indications that it
may be extremely effective in stabilising the surfaces of road
cuttings and similar places until such time as a plant cover,
whose growth is encouraged by krilium treatment of the
ground, can be established.

Weeds

Whereas in Britain weeds are one of the arable farmer's most
pernicious pests, in other lands they are coming to be regarded
as something like allies for the protection they afford the soil
against erosion. There was a time when a weed-free rubber
plantation was the pride of every plantation manager, nowadays a manager who did not allow weeds to grow abundantly
to protect his soil might risk losing his job. In the 'forestry'
system of rubber growing the jungle undergrowth is allowed
to develop, the soil is protected and yields of latex are not
reduced. The system is now almost universal in the Far East.
In tea plantations the position is rather more difficult, because
many natural weeds compete seriously with tea bushes for
moisture and nutrients. Systems of selective weeding are
being worked out in which those weeds which do more good
(soil protection) than harm are left, and those which do more
harm than good are destroyed.

Fertilisers

Erosion can usually be checked once a permanent plant cover
has been established, but plants will not thrive where the topsoil has gone or where soil fertility is already at a very low
level. In such cases fertilisers can be of great direct assistance
in erosion control by providing the necessary nutrients to enable plants to grow and hold the soil until erosion resistance
can be built up. The manufacture and utilisation of phos-

phatic fertilisers has been a major feature of the comprehensive soil-conservation programme carried out by the Tennessee Valley Authority. It has been reported from Russia that barren eroded lands have been restored to fertility merely by application of sulphate of ammonia which enabled grass to colonise the soil and subsequently form humus and produce a crumb structure. With the development of aerial application, especially in New Zealand, there is great scope for using fertilisers over wide areas for increasing the productivity and erosion resistance of otherwise inaccessible hill pastures.

But although soil-conservation measures are usually very simple and straightforward, the path of the soil conservationist who has to get them applied is beset with troubles. It does not take long to master the technical side of soil conservation. Difficulties arise when somebody in authority has to co-ordinate many different measures being carried out by many different people at many different times. He finds that conditions of land tenure, social and religious customs and economic questions are far more important factors in soil conservation than the length and steepness of the land slope, the nature of the soil, the intensity of the rainfall, the width of the contour strip and so on. Almost anybody could cope with the latter, almost nobody with the former. Where radical soil conservation is badly needed it may require an agricultural and social revolution to carry it through, comparable perhaps to that which helped to convert Britain from an agricultural to an industrial country. Agricultural revolutions do not happen so easily as political ones. They tend to be long drawn out and are seldom violent, but they need powerful stimuli to set them going, and soil erosion has been such a stimulus. It is for this reason that I suggest that soil erosion may be regarded by posterity, though not by this generation, as one of the greatest blessings that European civilisation has given to the world.

CHAPTER XIII

THE HISTORY OF BRITISH SOILS

THE soils of Britain, like those of most other countries, have always been very variable. It is impossible to speak of a 'typical' British soil, but we can speak of a British soil 'type'; that is to say, there is no real soil typical of the whole country, but there is a type to which the majority of the soils originally conformed more or less closely. This type is—or was, because it is now of rare occurrence—the brown forest soil formed under the deciduous forest which was the natural climax vegetation of much of the country. The main and most exciting event in the history of British soils since they became brown forest soils has been their conversion to agricultural soils. It took about a thousand years for the conversion to spread and be completed over the greater part of Britain. It could be done in a fraction of the time with the facilities and knowledge at our disposal today, but that knowledge is applicable mainly to brown forest and similar soils only. It is not applicable, for example, to tropical soils; when modern methods and techniques were applied to the rapid conversion of the soils of the groundnut areas of Tanganyika into agricultural soils the result was a disastrous failure.

The history of agriculture in this country is quite closely connected with the conversion of the soils from the brown forest type to the agricultural type. We shall try to trace this connection and to show how certain major stages in agricultural history influenced the gradual and imperceptible conversion of a forest soil type to a type which has shown by its performance that it is as well suited to intensive agriculture as any in the world. The main events of British agricultural history are described in two other books in this series—*A History of English Farming*, by C. S. Orwin, and *A History of Scottish*

Farming, by T. Bedford Franklin. I shall have to confine myself here chiefly to the history of English soils with which I am the more familiar, but the fundamental relationship between the conversion of English and of Scottish soils to agriculture and the history of agriculture is similar.

The main events of agricultural history, so far as they concern the conversion of the soil, are :

(1) open-field farming, based on the feudal manor ;
(2) the Black Death ;
(3) the rise of the wool trade ;
(4) the Tudor inclosures ;
(5) the introduction of clovers and roots, and the invention of the seed drill ;
(6) land reclamation ;
(7) the Industrial Revolution, leading to further inclosures ;
(8) the invention of artificial fertilisers.

These events were not necessarily consecutive, nevertheless they began roughly in the order given. Some, like open-field farming, lasted for centuries, whereas the Black Death was over in a few years, though its effects were felt for longer. Historically, of course, the events cannot be discussed outside the context of history, but for simplicity's sake each one can be regarded as having a quite definite influence on the course of the soil's evolution. Before we consider what these influences were, let us compare the properties of a typical agricultural soil with those of a typical brown forest soil.

By a typical agricultural soil we mean a soil on which intensive agriculture, including therein both arable and livestock farming, can be economically carried on. Such a definition will not satisfy everybody, but it is good enough. Few, if any, natural soils will support intensive agriculture for long. Soils have to be changed, and made more productive, for agriculture by man's activity which usually has to be applied continuously. The most important and typical properties of an agricultural soil are a well-developed crumb structure, adequate depth, a high content of plant nutrients in the surface layers, good drainage and an absence of excessive acidity or alkalinity—that is, the soil is one in which any ordinary farm crop, including grass, will do well. Any or all of these properties may disappear rapidly if man ceases to take proper measures to maintain them.

Note that nothing is said about texture, which is often the first property by which a farmer judges the quality of his soil. Difference in texture is largely a matter of the ease with which the conversion to an agricultural soil can be effected.

The brown forest soil displays to a considerable degree the essential properties of an agricultural soil, but these properties are maintained by the living forest, the destruction of which is the first step in the conversion of the soil to an agricultural soil. The history of agriculture in Britain has been the story of the discovery by man of ways of replacing and improving on the influence of oak and ash forest on soil evolution by his own activities.

The surface layers of a brown forest soil consist of well-decomposed humus and mineral soil which have been intimately mixed into mull by the action of the earthworms and insects that abound in this type of soil. The mull is usually only a few inches thick, but there is no sharp dividing line between it and the mineral soil immediately below. The amount of humus in the soil decreases with depth. The black colour of the mull blends into a cocoa-brown which gives the soil its name and merges gradually into the colour of the underlying rock where the soil proper ends.

The mull layer forms an excellent habitat for all kinds of moisture- and shade-loving woodland plants. Its fine crumb structure is produced by the intimate incorporation of the woodland humus with the soil, and by great microbiological activity in decomposing the humus and thus keeping the soil ' puffed up.' As the humus supply falls off in the subsurface layers the soil assumes a more cloddy structure more suitable for the coarse roots of trees than for the finer roots of flowering plants and herbs. The lower layers, however, contain more lime, potash and phosphate than does the surface, since these essential plant foods, needed in great quantities by human-food crops, tend to be washed downwards in the humid deciduous-forest climate—especially under cultivation when the loss from the surface by leaching is not offset by a compensating return of minerals in fallen leaves.

The large quantity of plant-food minerals that is required and taken up by deciduous-forest vegetation, and afterwards returned to the soil, is the principal biological factor in maintaining the stability of the brown forest soil. Without this continuous and efficient cycle of minerals from the lower layers through the plants and back to the surface of the soil, the humid climate would progressively impoverish and acidify the upper layers. Removal of the forest stops the main cycle of plant food from the root zone in the soil through the trees and back to the soil, but does not prevent the continuous washing-down under the influence of the humid climate. Hence deforested soils in a humid climate tend to become progressively poorer in plant food in the surface layers, apart altogether from any removal thereof in harvested crops. They also become poorer in humus when deprived of the annual leaf fall and tend to lose the original crumb structure of the mull layer.

These effects of deforestation are universal and well known. They are the reasons for the practice of shifting cultivation in nearly all primitive societies living on former forest soils. Cultivation results in rapid loss of soil fertility unless measures beyond the capacity of primitive peoples are taken to prevent it. Settled agriculture becomes possible only when the people have acquired sufficient social organisation for the community to exercise control over the use made of the land, and thus to retard, if it cannot actually prevent or reverse, the progressive exhaustion of the soil.

In Britain the social organisation that developed in the early days of forest-soil agriculture was based on the feudal manor and permitted only one system of land utilisation that was designed, firstly, to keep the forest at bay, and secondly, to slow down the inevitable exhaustion of the soil under agriculture. The system was known as the two- or three-field system, and comprised an invariable rotation, continuing for centuries, of either a winter-grain crop and a fallow or a winter-grain crop followed by a spring-sown crop followed by a fallow. The unchanging food crops were essential and only just sufficient to keep the people alive, and the fallow, even though it occupied a half or a third of the potential food-producing land, was absolutely indispensable to rest the soil and save it from rapid and complete exhaustion.

This type of agriculture, which survived in some parts of Britain into the nineteenth century, did little towards converting the brown forest soil into an agricultural soil. The virgin brown forest soil, indeed, had many of the properties of an agricultural soil, *but it could not keep them under the agriculture of the feudal age.* What feudal agriculture did for the soil was to change it away from the desired agricultural type, but in an orderly way which allowed agriculture to be continued while the soil deteriorated. There was no scope for improving the soil under the conditions of feudal agriculture, but equally there was no scope for that complete destruction of the soil by erosion that has been so frequent a consequence of the modern exploitation of new land by undeveloped societies. The soil slowly but inevitably lost its structure, its humus and its plant-nutrient reserves. Crop yields probably fell ; the yield of wheat in the fifteenth century was about 10 bushels per acre.

A feature of early brown-forest-soil agriculture was the use of the mould-board plough, which, instead of merely scratching the soil as earlier ploughs had done, turned it right over so that the underlying soil was brought to the surface. Thus the mould-board plough helped—like the forest, but less efficiently—to prevent the washing-out of plant nutrients from the soil by rain. It also produced the ridge-and-furrow effect which was so characteristic of arable land in England and served an important purpose in drainage. The ridges and furrows can still be seen in some old pastures which have been undisturbed since the land was inclosed, but were formerly cultivated. The ridges and furrows usually conform to the natural drainage of the field. Crops were planted on the ridges which were the driest part of the field, and excess water was drained away in the furrows.

Drainage was an often essential process in the conversion of forest soils to agricultural soils. The most important cultivated crop—wheat—of temperate agriculture is a plant adapted to a drier climate than that of most of Britain, and getting rid of the excess of water has always been a problem which is still not completely solved. The open-field farmers made a good start, and since the configuration of the open fields is preserved in the lay-out of many present-day farms, their influence on the conversion of the forest soils still persists. Later, when it was possible to extend the cultivated area by reclamation, many soils, such as fens and marshy soils, which originally were entirely different from agricultural soils, were successfully drained, and in the nineteenth century the development of under-drainage widened further the possibilities of reclamation.

Such progress, however, belonged to a later age. In the Middle Ages drainage operations were mostly confined to removal of surface water by making the best use of the natural lie of the land, and in particular by ploughing furrows up and down slopes. (Note that in many new countries, where people are today trying to convert other types of soil into agricultural soils, up-and-down ploughing has been found to be most harmful as it leads to soil erosion, and contour ploughing across the slope is advocated, and sometimes enforced, to conserve water and soil.) The drainage of the Middle Ages did little beyond making agriculture possible on soils which had lost much of the permeability and porosity they had possessed under the original forest.

In the Middle Ages the only way of getting additional plant food into the soil to compensate for that removed in crops was by applying animal manure. There is no evidence that medieval Britons systematically returned their own excreta to the land, as is done in some parts of China today. The lord of the manor often had special rights to animal manure (e.g. sheep had to be folded on the demesne lands at night), and as livestock numbers were low the fields of the common people can never have got enough manure to compensate for the loss of plant food in crops. One of the benefits that inclosure conferred on the soil was that it enabled more livestock to be kept, and hence more manure to be provided. This, however, by itself did not, and obviously could not, increase the fertility of the soil. All it did was to reduce the rate of loss of plant food that is greater when farm produce is fed directly to humans who do not return their excreta to the soil than when it is fed indirectly through animals which do. The actual reversal of the loss of plant food, by both cropping and washing out, that had gone on for centuries was not accomplished until the invention of artificial fertilisers in the nineteenth century.

Open-field agriculture might have broken down under any circumstances, through soil exhaustion, but other forces operated which destroyed the social foundation on which the system was built. Among these factors were the growth of towns and a money economy, the absence of many great land-lords at the Crusades, and the Black Death which administered the *coup de grâce* to the feudal system. The general effects of feudal agriculture on the soil had been to destroy much of the crumb structure that had developed under the former forest cover and was so essential to agriculture, and to deplete the soil of its plant-food reserves. The land was becoming very dirty, and weeds, the vanguard in the reconquest of the land by the forest, were gaining the upper hand.

In actual fact, of course, the soils of Britain at the end of the feudal age were in all stages of fertility, from highly fertile to completely exhausted—as they are today. But today the most infertile land is usually that which has not been continuously farmed, whereas in the fifteenth century it was long-cultivated farm land. Farming nowadays promotes soil fertility in Britain, then it destroyed it.

The Black Death in 1349 virtually eliminated serfdom from England, and with it the compulsory services on the lord's land that, like the work-days worked on the collective farm by the modern Russian labourer, removed all personal incentive to better farming. A more serious obstacle to improved farming and enhanced soil fertility were the rigid rules of cropping on which the communal open-field system had to operate. The rules made it certain that soil fertility would decline gradually, but they also ensured that it would decline only gradually.

Roughly contemporaneous with the breakdown in the feudal system was the rise to pre-eminence of the wool trade. These two events between them had a profound influence on the agricultural development of the soil. Hereafter the effect of agriculture was to raise instead of, as hitherto, to lower the fertility of the soil. The breakdown of the feudal system and the emergence of a capitalistic economy led to widespread inclosures of the open fields. As a result of the high cost of labour for cultivation after the Black Death and the boom in wool, nearly all the inclosures made in the fourteenth and fifteenth centuries were of arable land which was forthwith put into sheep pasture by its new owners, who for the first time could do what they liked with it, instead of being tied by the immutable rules of communal farming.

Inclosure thus introduced an entirely new influence into the evolution of soil in human occupation. The old open-field agriculture had been unable to prevent a gradual decline in the fertility which the soil had originally possessed. The inclosures gave the soil a rest—and the most effective kind of rest that it could have had in the circumstances prevailing five hundred years ago. If manures and fertilisers are not available there is nothing like a grass cover for restoring productivity. The effect of grass on soil structure has been described in Chapter X, but that was not understood until a few years ago. Inclosures were made because it paid to put land down to permanent grass, and because a stratum of society was emerging that could make inclosures against the will of most of those who occupied the land.

The general effect on the soil of this first major movement towards inclosure was to begin to restore the physical fertility of the soil, by resting the soil and producing a crumb structure under grass. The effect spread very slowly through the country. Orwin states that less than a million acres were inclosed during the Tudor period, mostly in the English Midlands, but inclosure in pasture continued to be made until the end of the eighteenth century when high prices for grain and land rents shifted the emphasis from pastoral to arable farming.

Resting the land, however, or letting it go down to grass cannot be described as agricultural progress, because food production from pastoral is much less than from arable land of comparable productive capacity. Russian soil scientists believe that most of the good effects on soil structure of putting land down to grass are achieved in the first two or three years, and that they disappear in about the same time when the land is ploughed and cropped. According to Russian ideas, nothing is gained by keeping land in grass for more than a few years. Experience in this country, however, is that it may take up to twenty-five years to build up a stable crumb structure in old arable soil by putting it in grass (Low, 1951). Thus the individual farmer on his inclosed pastures was a hoarder of soil fertility, and the communal farmer on his open fields was a consumer of soil fertility. By mixing pastoral and arable farming a way was found of spreading the benefits of the former into the latter.

This represented a second phase in the formation of an agricultural soil. In Scotland and Wales the mixing took the form of ley farming, the principle of which is that land is alternately grassed and grazed to store up fertility, and cropped to use it up. In England it took the form of what is known as mixed farming in which temporary or permanent arable and pasture are found on the same farm and are worked together as an economic unit.

The first step was the introduction of clover and new and improved grasses after the spring-sown crop in the three-course rotation. These increased the output of hay, and allowed more livestock to be kept and more manure to be produced. The clovers themselves were valuable fertilisers, though their significance was not understood until late in the nineteenth century. They provided a means, by their nitrogen-fixing properties, of increasing the total amount of plant food in the soil without adding manure obtained by extracting plant food from the same or some other piece of soil. Clovers had been valued in Roman agriculture of the pre-Christian era for their soil-enriching qualities, and their introduction into western European agriculture represents the rediscovery, in a new environment, of a forgotten but well-tried principle of making soil fertility. The better swards obtained with the new grasses may also have helped to improve soil structure, but we have no evidence that they actually did.

The appearance of turnips as a major agricultural crop in the eighteenth century had a profound and beneficial influence on British agriculture, but its *direct* influence on the evolution of soil was not as great as that of the growing of clovers and grasses. Turnips and other root crops enabled a profitable use to be made of the fallow, and they solved the problem of winter keep of livestock, and thus made possible mixed farming as it is known today. But their cultivation did not have much effect on soil properties. Roots enabled the farmer to feed his animals in winter off the produce of his arable fields, and at the same time to grow more crops, carry more stock and to produce enough manure to retard the exhaustion of the soil under the plough. Later on, in the twentieth century, when most of the farm animals in Britain were fed on imported fodder and consequently yielded far more plant food in their manure than they took from British soils, the livestock industry became a positive factor in increasing soil fertility.

Another factor, operative in the seventeenth century, in the conversion to agricultural soils was the improvement in methods of cultivation. One of the purposes of cultivation is to keep the soil open and porous. Cultivation, by itself, cannot produce a crumb structure, but it can produce, if only temporarily, some of the qualities of a structural soil. In the days when all cultivation had to cease after the seed had been broadcast, the compaction of the soil by rain during the growing season, caking in dry periods and the unrestricted growth of weeds acted as serious and sometimes fatal checks on productivity.

It was Jethro Tull, the inventor of the seed drill, who discovered how to maintain some semblance of a soil structure while the crop was growing. Drilling seed in rows instead of broadcasting allowed cultivation to be done between the rows after sowing. Tull argued that ' when crops were sown, nature at once began to undo the effect of previous ploughings and sowings. The earth united, coalesced and consolidated, and so shut out the air and water from the roots, and decreased the food supply at the moment when the growing plants most needed increased nourishment. To some extent the use of farmyard manure kept the land friable ; but it also stimulated the growth of weeds. The better course, therefore, was to keep the land pulverised by tillage and so prevent the contraction of the food area of the growing crops. So long as wheat and turnips were sown broadcast, this method could not be satisfactorily employed. In two ways the crops benefited by constant tillage. In the first place the land was kept free from weeds, and so saved from exhaustion. In the second place the repeated pulverisation of the soil admitted air, rainwater and dews to the roots of the plants, and extended the range from which their lateral growths drew their food supplies.' [1] In short, the transient structure produced by cultivation and all the benefits it conferred on the crop could be maintained throughout the growing period by drilling instead of broadcasting the seed. Drill husbandry obviated the necessity of a resting fallow, formerly the sole means of maintaining arable land in good heart. By frequent cultivation Tull was able to grow heavy crops of wheat for thirteen successive years—an almost unheard-of achievement.

By the eighteenth century, therefore, men had learnt enough about the management of the former forest soils of Britain to be able to cultivate and crop them continuously and at the same time to maintain their fertility. So far as the soil was concerned, men were treating it as well as the forest had ; they kept it fertile and in good condition while they made it produce the plants they needed. The difference between the human and forest régime was that in the latter soil fertility was maintained by the action of the whole plant community, and particularly the trees, in the former it was maintained partly by the action of some of the plants, particularly clovers and grasses, but mainly by the physical actions of the cultivators. Average wheat yields, according to Arthur Young's writings, seem to have been in the neighbourhood of 20 bushels per acre, or double what they had been three centuries previously. The reinvasion of cultivated land by the forest had ceased to be a menace, though the forest came—and still comes—back quickly enough if land were abandoned.

Having at last mastered the art of producing an agricultural soil, British farmers were in a position greatly to extend the area of their conquest on to hitherto uncultivable land. In the eighteenth and nineteenth centuries large investments were made in the reclamation of waste lands—many being sound economic propositions, some, if judged by orthodox rules, unsound. Among the latter propositions may be included

[1] *English Farming*, by Lord Ernle

the reclamation of Exmoor forest and Scottish heathlands at great cost in money or labour, or both. Nevertheless, the reclaimed land has been a productive asset to the nation over the ensuing centuries, and has extended the area of man's precarious dominion over Nature. Where hitherto it had taken a thousand years to convert a forest soil into a soil fit for permanent agriculture the process could now be completed in a decade or two.

The main reason for this phenomenal acceleration of the process of making an agricultural soil was not so much the actual progress made in the arts of agriculture as the changes in the economic environment that made the progress possible. Most of the agricultural improvements had to wait until communal tenure of land had been replaced by individual tenure, and the individual landowner had freedom to do what he liked with his property. The early step in soil conversion, of resting the soil under grass, could have been taken at any time if there had been land available for resting and if custom had allowed it to be rested. Under communal tenure there was no incentive to find ways of improving the soil, and the soil was not improved. Throughout history most advances in soil management have been the result of applying common sense, and could have been made at any time if the incentive to make them had been there. Thus recent advances in the art of soil conservation are based on principles that have been known for thousands of years, but it needed the incentive of soil erosion to get them applied.

In the latter half of the eighteenth and in the nineteenth centuries another potent factor in getting progressive agricultural measures applied was the availability of money made in industry and commerce for investment in land improvement. Such money, for example, reclaimed Exmoor, and many other costly land-reclamation schemes were carried through by industrialists who saw in land improvement either a safe investment for their wealth or a means of raising their social prestige. This benefit, of manuring land with money,

conferred on the soil by the Industrial Revolution was associated with the growing need to expand the cultivated area
to feed the rapidly increasing urban populations. There was
a second spate of inclosures, mostly of commons, waste land
and woodland, and this time—unlike the Tudor inclosures—
for cultivation, not for pasture, that did not cease until practically all the cultivable land in the country had been inclosed.
In Tudor times inclosure for cultivation would have had a
baneful effect on soil fertility, but three centuries later land
could be quickly enriched by cultivation. Crop rotations and
improved farming techniques played an essential part in
making this possible, but more fundamental were the social
and economic conditions without which the agricultural
improvements could not have been applied. Inclosure of land
and its tenure in severalty were obvious factors facilitating
improvements, but equally significant was the landlord-tenant
relationship which developed over much of Britain in the
nineteenth century. It enabled large sums of money earned
in industry to be invested in land improvement by wealthy
owners, while the actual farming was done by tenants operating
relatively small areas to which they could give the individual
attention required by the intensive high farming of the nineteenth century. The evolution of soil under agriculture—
whether the soil is enriched or impoverished thereby—depends
just as much on the kind of human society colonising it, as
does the evolution of a ' natural ' soil on the kind of plant
society colonising it.

But the event which had the most dramatic effect on
agriculture was the invention of artificial fertilisers. Until
artificials appeared it was almost impossible to enrich a soil
with mineral plant nutrients without depriving another soil
of them. The only manure of consequence was the dung
of animals which had already consumed food containing plant
nutrients taken out of the same or a neighbouring soil. Animal
manure tended to enrich the soil in humus which may have
indirectly enriched the soil in plant nutrients by stimulating

biological activity and accelerating rock weathering, thereby bringing mineral plant nutrients out of the unweathered rock and into the biological cycle. Some soils were definitely enriched by the increasing amounts of dung made available by intensified farming, but in general the additional nutrients received were got at the expense of pastures. The consequent impoverishment of some of these was not revealed until they were ploughed up in the Second World War.

In the humid climate of Britain there is a continual washing-out from the soil of all plant nutrients, especially nitrogen, calcium and potassium. To this must be added the unavoidable removal of plant nutrients in harvested crops. Such losses can be kept low, but cannot be altogether avoided, by returning to the soil as many plant residues, in the form of organic manures, as possible. Under virgin forest the loss is confined to the very small amounts of nutrients that are not absorbed by plant roots and are leached out of the soil by rain. In British conditions it would take centuries for the effect to become apparent in the soil's economy. Under agriculture the loss is very much greater, and falls particularly heavily on calcium which, in the form of lime, is the element which chiefly prevents a soil from becoming acid.

The commercial production of nitrogen, phosphate and potassium fertilisers since the second half of the nineteenth century has made it possible greatly to reduce, but not yet to reverse, the total loss of plant nutrients from British soils, and at the same time greatly to increase crop yields. Thus average wheat yields have risen from about 10 bushels per acre in 1500 to 20 bushels in 1830, to 30 bushels in 1900 and to 38 bushels in 1950, and the tendency is still upwards as also is the consumption of fertilisers. Nitrogen added in fertilisers is either immediately taken up by the crop or washed out of the soil, so there is no cumulative gain of nitrogen by the soil; added phosphates and potash are partly absorbed by crops and partly 'fixed' in the soil in chemical forms which are unavailable to plants. Some of these fixed nutrients seem to become available in subsequent years and to have a ' residual ' fertilising effect, nevertheless it is astonishing how much phosphate, only a fraction of which is taken up by crops and little of which is washed out of the soil, can be poured into some soils year after year and still both crops and soil remain hungry for more. At present we know very little about, and have very little control over, what happens to the so-called artificial fertilisers when they are added to soil, and it is possible that they will not have any great enduring effect on soil economy, but their short-term year-to-year effects in increasing crop yields are obvious enough. By the one criterion—production—of soil fertility that counts, the fertility of British soils, has doubled in the last hundred years and is still going up. The fact that production would fall next year with a bump if fertilisers were not used is no reason for calling fertilisers a dangerous stimulant ; there would be the same result if any other essential farming operation were omitted.

In 1949 some 200,000 tons each of nitrogen (N) and potash (K_2O) and 400,000 tons of phosphoric acid (P_2O_5) were applied to British soils. Not more than half the nitrogen and a good deal less (a quarter to a fifth) of the potash and phosphoric acid were recovered in crops, but even with those low efficiencies, which are a good deal higher than are obtained from organic manures, the use of fertilisers was generally very profitable. Markedly to increase the efficiency of fertilisers would represent as great an advance in agricultural technique as did their invention.

The use of fertilisers raised soil productivity not only directly by supplying additional amounts of plant nutrients immediately available to crops, but also indirectly by providing farmers with a powerful incentive to keep their land in good heart. Fertilisers were largely wasted on dirty or badly drained land. The profit from them was greatest when the physical condition of the soil was the best possible for crop growth. Although the nineteenth-century farmer did

not think about soil structure in the modern jargon, he discovered that certain operations (which did promote structure
formation) were necessary to enable him to get the greatest
return on his outlay on fertilisers. In particular, the profits
obtainable from fertilisers on land in good heart induced
farmers to apply the new and costly practice of under-drainage
to much land which, without drainage, could never have been
brought into the structural condition characteristic of an
agricultural soil.

A degenerative chemical process set in motion in the soil,
by the cutting-down of the original forest, that agriculture has
not yet been able to check completely, is the washing-out of
lime or, more correctly, of calcium. Calcium, though essential
to plant growth, is not required in such large quantities as
nitrogen, phosphorus or potassium, and any surplus available
in the soil is liable to be lost. It has been estimated that under
average British conditions a half to one hundredweight of lime
(CaO) per acre is lost in this way every year. Loss of lime,
unless checked, results in acidity which will in time render the
soil unsuitable for agricultural crops. Fortunately there is no
lack of chalk and limestone in Britain, but vast quantities are
needed by our soils every year, not only to make up the annual
loss in drainage waters, but also to restore the greater losses
accumulated over many centuries when little liming was done.
Much of the grassland ploughed up in the Second World War
was found to be very deficient, for cropping purposes, in lime
which had been lost in drainage waters, and in phosphate
which had been lost in milk and hay. Since 1937 liming has
been officially encouraged by a State subsidy which has im-
pressed the benefits to be gained from the practice on a steadily
widening circle of farmers.

Let us now briefly recapitulate the history of British soils
under agriculture—what they were like before they were
cultivated, what they suffered under feudal farming, and how
they have benefited under recent farming which for our purpose may perhaps be literally described as capitalistic because

the benefits enjoyed by the soil were mostly derived from heavy manuring with capital.

The original forest soils were in a state of equilibrium with the vegetation and the climate. They did not change appreciably from one century to the next. The rate of disappearance of old humus was equalled by the rate of formation of new humus from dead plant residues. The soil fauna and microorganisms were just adequate to perform their parts in the processes of humification. Plant nutrients absorbed by the vegetation were equal to those returned to the soil in dead residues. There was probably a slow leakage of nutrient and other elements in drainage waters, roughly equivalent to what was released from unweathered rock by weathering, but this would not become apparent in the properties of the soil over a thousand years. The general biological conditions ensured a fairly even distribution of mull humus down to the parent rock, a loose crumb structure on the surface and a sufficiency of all plant nutrients to support an active vegetation.

Such soils were very suitable for agriculture, but lost their good qualities quickly under cultivation. Plant nutrients were liable to be washed out, particularly during the winter when nothing was growing on the soil, and some were removed from the soil in crops, the soil was exposed directly to the beating action of rain, lost its structure, and tended to become more compacted and less pervious and porous. The arts of agriculture were directed towards preserving the good original properties of the soil for as long as possible. The most important measure to this end was to rest the soil in fallow every second or third year. Other measures included the use of the mould-board plough to turn the soil over and thus to keep plant nutrients from being washed out of the ploughed layer, and laying out the fields to secure the greatest surface drainage. Cultivation also gave the soil a temporary looseness which was a poor substitute for the natural soil structure developed under forest. Whatever manures were available cannot have been sufficient to compensate even for the loss of plant nutrients removed in the harvests. The measures were insufficient to prevent soil deterioration, let alone to effect any soil improvement, but they did make it possible to keep land under cultivation, sometimes for centuries. Had the British brown forest soils been treated in the same way as were the much richer virgin prairie soils of North America in the nineteenth century they would have given out in a few years.

Thus the effect of substituting medieval agriculture for a deciduous forest, or, indeed, any other type of vegetation was to produce a hungry, structureless, acid and ill-drained kind of soil. Note that it was not the medievalism of agriculture that caused these unfortunate effects, for modern agriculture has had equally unfortunate effects in other climes, but the maladaption of agriculture to the environment. When a social economy had evolved that made it profitable to conserve and enhance soil fertility, the potentialities of the environment were exploited to the full without anybody planning it.

With the appearance of a capitalistic agriculture—which for convenience we can date from the time of the Tudor inclosures—conditions arose which favoured the formation of an agricultural type of soil. The first step in this direction was the prolonged rest the soils got under grass. Such a necessary rest was at that time only possible because certain capitalistically minded people, as they would be called nowadays, had gained the economic power to inclose the land to their own and the land's advantage and to the detriment of the ordinary peasant who lost his land. Once inclosure had been effected and the occupier of the land was no longer tied by the restrictions imposed by communal tenure, many agricultural practices were introduced that had a very beneficial influence on agricultural-soil development—the growing of clovers and grasses, drill husbandry and improved cultivation, the adoption of crop rotations, under-drainage and the use of fertilisers and lime. The result of all these and other practices has been that crops which by nineteenth-century standards would have been regarded as enormous, are now taken from the soil year after year, and productivity continues to rise.

This is a measure of the triumph of agriculture over the forest. Five hundred years ago, one or possibly two good crops could have been expected off newly cleared forest land, then it would have been necessary to rest the soil, and probably crops as good as the first would never again be obtained. Today, the best crops of five hundred years ago would be considered scarcely worth harvesting, but almost any land can be made to produce and go on producing huge crops within a few years provided enough money is spent on it. In the tropics today agriculture is in much the same stage relative to the environment as it was in Britain five hundred years ago. One or two crops can be taken off cleared land which must then be abandoned. Occasionally and increasingly permanent agriculture is practised (as it was in Britain five hundred years ago), but in general the economics of tropical agriculture do not permit the necessary adjustments to be made that would make agriculture a soil-conserving operation. The use of fertilisers, for example, is an obvious way in which soil fertility could be conserved in colonial countries whose economic progress depends on the development of an export agriculture, but hitherto it has seldom paid to fertilise extensively.

The typical agricultural soil, which exists only in approximation, of Britain of the second half of the twentieth century has many of the properties of the ideal, man-made agricultural soil. It is sufficiently deep and can be further deepened if required, it has a stable crumb structure which is maintained as necessary by manuring or a spell under grass, it is recovering from the acidity produced by earlier exhaustive farming in a humid climate, it can have as much plant food as it wants, and it is or can be adequately drained. Chemically and physically it is not markedly different from the original soil, but biologically it is different because it supports a human population instead of trees, as well as a ground flora quite

different from that of a forest. Besides, continued human
activity is necessary to preserve its properties. If human
influence were withdrawn the qualities of an agricultural soil
would disappear as quickly as would the qualities of a forest
soil if the trees were cut down.

The conversion of British soils to agriculture has been a
long and tedious process, for the accomplishment of which
nobody should take much credit, for those who accomplished
it did so in response to ordinary economic stimuli. The soils
are still far from being ideal for agriculture, and far from being
as productive as they might be. It is impossible to forecast
how much further soil productivity will rise. Yields of 80
bushels of wheat per acre are not unknown, and there is
enough science available to make such yields common, but
it is doubtful whether there is enough money or inducement
to do so. It would be unwise to predict whether or not there
ever will be. As the second half of the twentieth century opens,
trends are apparent towards a greater use of fertilisers and lime,
and an extension in England of the kinds of alternate husbandry
or ley farming that have for long been common in Scotland.
The trend towards ley farming has reduced the area under permanent grass and increased the areas under temporary grass
and under the plough. This trend acts in an opposite direction
to the trend towards specialised farming that has been apparent
in the last fifty years. It tends to mix arable and livestock
farming. The effect on the soil of an era in which ley farming
became general might be similar to that of the early inclosures
for pasture. It should help to preserve the crumb structure
of the soil while using the land alternately for pasture and
tillage, but we cannot yet say definitely that ley farming does
have this effect. It is a difficult question to answer experimentally. A complicated experiment has recently been
started at Woburn in Bedfordshire to discover exactly what
a ley does to soil fertility, but full results will not be available
for many years. First results indicate an increased yield of
about $1\frac{1}{2}$ tons on a total yield of 13 tons of potatoes per acre

as a result of a three-year grazed ley, as compared with three years of arable rotation.

But the future of British agriculture will not be determined, although it may be guided, by the results of scientific experiments. It will be determined by economic and social influences, and principally by the compelling need for an impoverished, densely populated country to produce not only most of the food but also more of the timber it consumes. Politicians and economists are continually impressing on us the need to strain every effort to produce more coal, more steel and more nylons to sell abroad for food that could be produced, with less effort though perhaps more money at first, from millions of derelict acres which now yield little or nothing, and most other acres which could yield more than they do. The soil, well managed, is an almost inexhaustible source of the first raw material required by every industry and profession. For the short way ahead that can be seen there seems to be no alternative for the British people, if they are to survive as a nation of standing, but to exploit to the utmost their greatest but still only half-used resource, the land. Given favourable economic conditions and a determination, which is at present half-hearted, on the part of the nation's administrators to get the job done, it can be done and a high level of soil fertility maintained indefinitely. Most of the required knowledge is already at our disposal, soil science is still in its infancy, and great advances may be expected in the near future. But whether or not the job is done will ultimately depend on political and economic events at home and abroad, and soil scientists have little influence on those events. Whatever happens we can rest assured that they, at any rate, will never become our masters.

BIBLIOGRAPHY

of books and papers referred to in the text

ABLEITER, J. K.   *Missouri Agric. Expt. Sta. Bull. 421,* 13–24 (1940)
BEIJERINCK, M. W.   *Bot. Ztg. 46,* 725–35 ; 741–50 ; 757–71 (1888)
BORNEBUSCH, C. H.   *Forstl. Forsøgsv. Danmark 11,* 1–256 (1930)
CLARKE, G. B. and MARSHALL, T. J.   *Aust. J. Council Sci. Indust. Res. 20,* 162–75 (1947)
CLARKE, G. R.   *J. Soil Sci. 2,* 50–60 (1951)
COILE, T. S.   *J. Forestry 33,* 726–30 (1935)
CUMBERLAND, K. B.   *Soil Erosion in New Zealand* (Wellington, 1944)
DARWIN, CHARLES.   *The Formation of Vegetable Mould through the Action of Earthworms* (London, 1881)
DAVIES, W. M.   *Agric. Prog. 15,* 111–18 (1938)
DHAR, N. R. and MUKERJI, S. K.   *Proc. Nat. Acad. Sci. India 6,* 136–48 ; 289–303 (1936)
ENSMINGER, L. E. and GIESEKING, J. E.   *Soil Sci. 53,* 205–9 (1942)
EVANS, A. C.   *Ann. Appl. Biol. 35,* 1–13 (1948)
FALKENSTEIN, VOGEL VON.   *Int. Mitt. Bodenkunde,* 495–517 (1911)
FAULKNER, E. H.   *Plowman's Folly* (Norman, Oklahoma, 1943)
FULLER, W. H.   *Proc. Soil Sci. Soc. Amer. 1946, 11,* 280–3 (1947)
GARMAN, W. L.   *Proc. Okla. Acad. Sci. 28,* 89–100 (1948)
GLYNNE, MARY D. and MOORE, F. JOAN.   *Ann. Appl. Biol. 36,* 341–9 (1949)
HENDRICKSON, B. H., CARREKER, J. R. and ADAMS, W. E.   *Soil Conservation 9,* 138–41 (1943)
HOFFMANN, R. W.   In *Blanck's Handbuch der Bodenlehre 7,* 431 (1931)
HOPP, H., and SLATER, C. S.   *J. Agric. Res. 78,* 325–39 (1949)
JACOT, A. P.   *Sci. Mon. N.Y. 40,* 425 (1953)
KATZNELSON, H., LOCHHEAD, A. G. and TIMONIN, M. I.   *Bot. Rev. 14,* 543–87 (1948)
KEEN, B. A.   *J. Roy. Soc. Arts 90,* 545–79 (1942)
KRASILNIKOV, N. A.   *Mikrobiologiya 13,* 144–6 (1944)
KRASILNIKOV, N. A.   *Ibid. 18,* No. 2, 49–58 (1949)
KRAUSSE, A.   *Naturwiss. Wochenschr. 15,* 371 (1916)
KUBIENA, W. L.   *Micropedology* (Ames, Iowa, 1938)
LEEPER, G. W.   *Introduction to Soil Science* (Melbourne, 1948)
LIPMAN, J. G., and CONYBEARE, A. B.   *N.J. Agric. Expt. Sta. Bull. 607* (1936)
LOCHHEAD, A. G.   *Trans. Roy. Soc. Canada 42,* Sect. V, 73–80 (1948)
LOWDERMILK, W. C.   *Trans. 3rd Int. Cong. Soil Sci. 2,* 181–94 (1935)

MARTIN, W. S. *Emp. J. Expt. Agric. 12,* 21–32 (1944)

MELIN, E. *Trans. Brit. Mycol. Soc. 30,* 92–9 (1948)

MILNE, G. *Trans. 3rd Int. Cong. Soil Sci. 1,* 345–7 (1935)

MÜLLER, P. E. *Studien über die natürlichen Humusformen* (Berlin, 1887)

NĚMEC, A. *Forstarchiv* 497–503 (1929)

OVINGTON, J. D. *J. Ecol. 39,* 363–75 (1951)

PAVLYCHENKO, T. K. *Canada Natl. Res. Council, No. 1088* (1942)

RAYNER, M. C. *Ann. Appl. Biol. 30,* 397–9 (1943)

ROBINSON, G. W. *Soils. Their Origin, Constitution and Classification,* 3rd ed. (London, 1949)

ROSEAU, H. *C. R. Conf. Pédol. Méditerr. 1947,* 389–404 (1948)

RUSSELL, E. J. *Soil Conditions and Plant Growth,* 6th ed. (London, 1932)

RUSSELL, E. J. (revised by E. W. RUSSELL). *Soil Conditions and Plant Growth,* 8th ed. (London, 1950)

RUSSELL, E. J. and HUTCHINSON, H. B. *J. Agric. Sci. 3,* 111–44 (1909)

RUSSELL, E. W. *Rothamsted Expt. Sta. Rept. 1949,* 130–47 (1950)

RUSSELL, M. B. *Proc. Soil Sci. Soc. Amer. 1939, 3,* 51–4 (1940)

SCHOFIELD, R. K. *Trans. 3rd Int. Cong. Soil Sci. 2,* 37–48 (1935)

SEKERA, F. *Z. PflErnähr. Düng. 52,* 57–60 (1951)

SHOREY, E. C. and MARTIN, J. B. *J. Amer. Chem. Soc. 52,* 4907–15 (1930)

SMITH, G. *Organic Surface Cultivation* (London, 1950)

SOIL SURVEY OF GREAT BRITAIN. *Report* (1950)

SPINKS, J. W. T. and BARBER, S. A. *Sci. Agric. 28,* 79–87 (1948)

STAMP, L. D. *The Land of Britain and its Use* (London, 1948)

STORIE, R. E. *Calif. Agric. Expt. Sta. Bull.* 556 (1933)

VILENSKY, D. G. ' Aggregation of Soil, its Theory and Practical Application,' *Commonwealth Sci. Indust. Res. Org. Melbourne* (translated from Russian by A. Howard) (1949)

WAKSMAN, S. A. *Humus* (London, 1936)

WAKSMAN, S. A. and IYER, K. R. N. *Soil Sci. 34,* 43–69 (1932)

WAKSMAN, S. A. and IYER, K. R. N. *Ibid. 36,* 57–82 (1933)

WILDE, S. A. *Agron. J. 42,* 522 (1950)

WINTER, A. G. *Z. PflKrank. PflSchutz. 56,* 93–5 (1949)

INDEX

Acidity, soil 43–6; caused by fertilisers 41–2; caused by soil exhaustion 211; crop tolerance of 45, 142; earthworms and 69, 74, 142; 'exchange' 39; micro-organisms and 81; neutralised by lime 12, 134, 206

Actinomycetes 80, 101–2

Agricultural soil, development of 197, 203–4, 210–12; properties of 3, 16, 27, 147–51, 163, 193–4, 211

Algae 80, 98; nitrogen fixation by 91, 98

Alginic acid, structure formation by 108

Alkaline soils 26, 40, 45–6, 52, 91, 160

Alluvial soils 160

Analysis, soil 31, 33

Antibiotics 81–2

Ants 73–4

Association, soil 166, 168

Azonal soils 160–1

Azotobacter 90–1, 98

Bacteria 5, 75–6, 80, 82–4, 98, 115, 143–5, 148; competition with plants 87, 95, 97; nitrogen-fixing 88–92, 150; numbers in soils 83, 95

Bacterial gums, structure formation by 107, 109

Basic slag, use in forestry 139–40

Biological cycle 32, 48, 50, 84, 144, 206

Birch, effect on forest soils 134–5

Black Death 193, 199

'Boke of Husbandry,' quotation from 127–8

Boron-deficiency diseases 52

Broadbalk field, available nutrients in 36; humus in 78

Brown forest soil 11–12, 134, 153, 156, 158, 162, 192–3, 209

Capillarity 54–6, 61, 117, 145

Capillary fringe 62

Capillary moisture 54, 60–1, 117

Capillary potential 56–9

Capillary theory 117–19

Carbon-nitrogen ratio 86–7

Catena 168

Chernozem 13–15, 65, 120, 143, 147, 151, 153, 156, 162, 168

Chestnut-coloured soil 154, 156

Chlorophyll 51, 98

Classification of soils, American or binomial 163–8; by productivity 169–71; genetic 153–63; purpose of 162–3; Russian 152

Clay fraction 19

Clay-humus complex 39, 42, 70, 75, 102–3, 110, 113

Clay minerals 23–4, 98

Clay soils, crops suitable for 22; properties of 21–2

Clear cutting 137

Climate, influence on soil formation 7, 9–13, 147, 152–9; zones of 9, 156–8, 160

Clostridium 90

Cobalt-deficiency diseases 52

Colloids 19, 25–6, 32, 38–43, 113

Combined water 54

Compost 49, 81, 100, 102

Coniferous forest 6, 11, 134, 141; faunal activity in 71, 75–6; mycorrhizas in 94; nutrient requirements 136–7; regenerated by birch 134–5; soils 8, 10–11, 130–2, 142–3, 153

217

Printed in Great Britain by
Thomas Nelson and Sons Ltd, Edinburgh